# SIXTY YEARS
### (and a bit)
# WITH SMOKER AND VEIL

Robert N H Skilling

First published June 1991 by
Northern Bee Books,
Scout Bottom Farm,
Mytholmroyd,
Hebden Bridge,
West Yorkshire HX7 5JS

Book Design by Karen Sutcliffe

This edition re-published 2012

ISBN  978-1-908904-16-4

# SIXTY YEARS
## (and a bit)
# WITH SMOKER AND VEIL

Robert N H Skilling

**NORTHERN BEE BOOKS**
**Mytholmroyd, Hebden Bridge**

# Contents

Dedicated to my wife Jean who has been hostess to countless beekeepers throughout our forty years together.

# Chapter One

It is many years since my old mentor in beekeeping, the late Robert Brown of Stewarton impressed upon me a few simple rules for successful beekeeping:

(a) never try to impose your theories upon the bees - work with nature, not against it,
(b) never disturb the brood nest unless you have a very good reason for so doing - you must have a clear idea of what to do. Each disturbance of the brood nest costs you honey and may precipitate the building of swarm cells,
(c) aim to have the colony at maximum strength for the honey flow,
(d) try to keep the foraging strength of the colony intact.

There you have the foundations of my beekeeping philosophy. My friendship with Robert Brown began when I was a boy of fourteen years of age and continued through many years until he died in his eighties.

He was the most methodical beekeeper that I have ever known. He was a man of sterling worth whose knowledge and intelligence far exceeded his formal education.

But Robert Brown is not the only one to whom I am greatly indebted. Indeed it is doubtful if any one can claim to have much knowledge of bees, or beekeeping that is original to themselves. We all owe an immense amount to others.

The late Dr. John Anderson, one time Head of the Beekeeping, Department of the North of Scotland College of Agriculture, Aberdeen, had a tremendous influence on Scottish beekeeping in my younger days. He was for many years the distinguished Editor of The Scottish Beekeeper. Little did I think when I was devouring his writings in that journal, that one day I myself would fill his editorial chair and become the longest serving editor of it to date.

Dr. Anderson played a small part in the discovery of the cause of acarine disease (at the time known as the Isle of Wight disease) which practically wiped out the bee population of the

United Kingdom at the beginning of this century. He was a science master at the Nicholson Institute, Stornoway, at the time, and suggested that the cause of the disease could be mite infestation.

The actual discovery of the mite as the causative agent of the disease was made by Dr. Rennie at Aberdeen University, although the first person to actually see the mite was Dr. Rennie's assistant, Miss Elsie Harvey. The thrust of Dr. Anderson's teaching was that the way to successful honey production is to have large, powerful colonies - *"Surplus bees make surplus honey"* was his dictum. *"Success attends those who keep most bees in fewest hives"* was another of his sayings. This teaching greatly reinforced all that I had learned from Robert Brown.

Margaret (Peggy) Logan, an assistant of Dr. Anderson, enthusiastically advocated his teaching even years after his death. Of all the college lecturers I have known, Peggy Logan is the one who, by far, commanded my highest respect as a sound, practical beekeeper. I was early drawn into a group of beekeepers who followed the teachings of Dr. Anderson. Notable amongst these was Dr. James Tennant and his wife Frances of Glasgow, who for many years was a towering figure in the affairs of the Scottish Beekeepers' Association. Others all along my sixty years in beekeeping were to influence my beekeeping.

# Chapter Two

As is well known, the beekeeper's year begins in the autumn, so I may as well begin with that time. The removal of the final honey crop and the preparation of the colonies for winter are really all one operation. Bees are not so easily handled once they have realised that they have been dispossessed of their hard won stores so it is better to get all the major manipulations over as soon after the supers are removed as possible.

Before the clearing boards are put in place the beekeeper should satisfy himself (or herself) that the colony is queen right; that is that it is headed by a vigorous young queen and that there is abundant brood at all stages. He should also be satisfied that both bees and brood are healthy. The key to this is that every beekeeper should be thoroughly familiar with that which is normal in a strong, healthy colony. Anything that strikes the beekeeper as abnormal should be investigated. Unfortunately advisory services are no longer free to the beekeeper, but limited services are available at a very nominal fee.

The combs should be arranged for over-wintering. There should be two or three empty, or nearly empty combs in the middle of the deep chamber to provide clustering space for the bees. I normally put the permanent super, that is the super which has been above the deep brood box during the active season and which was available to the queen, under the deep chamber for over-wintering - unless, of course the colony has two deeps for a brood nest, but even in this case I reverse them in the autumn. This procedure has several advantages. First of all it provides a convenient passage-way for the bees, facilitating the easy movement of the cluster on to fresh stores without having to descend into the cold air near the entrance in frosty weather. I am convinced that this mid-passage has saved many a colony in periods of protracted frost.

Another advantage is that as spring advances it will be found that the bees and queen have moved up into the deep chamber where brood rearing is in progress. This permits easy spring management without serious disruption of the brood nest.

The beekeeper must also satisfy himself as to the adequacy of each colony's stores. He may now put on the clearing board and proceed with the harvesting of the honey crop. The extracted combs may be stored 'wet' or 'dry', according to the preference of the beekeeper. For the benefit of the uninitiated 'dry' combs are obtained by returning the extracted combs to the colony, above the clearing board, allowing the bees access to them by opening up a very small aperture. This involves removing the porter escape but closing up the feed hole leaving only just enough room for the bees to gain access to the wet combs. They will clean up the combs leaving them quite dry. Once the bees have deserted them the supers can be removed for storage.

They should be stacked with a queen excluder under the bottom one and another one over the top. These are effective in keeping out mice. Sheets of newspaper placed between supers, leaving no gaps will help to keep out wax moths which can devastate precious drawn comb. An alternative is to seal each super in a plastic bin bag. These must be completely sealed with adhesive tape to keep out all intruders and they should be stored in a dry place. Any feeding ought to be done at this time, for any feeding necessary during  winter - or even in the spring simply indicates that the beekeeper has failed to feed his colonies timorously and adequately. I find that a sugar solution suitable for autumn feed is obtained if the dry sugar is shaken level in whatever utensil is being used: the level carefully marked and boiling water is added until it reaches the level marked with the dry sugar. The sugar must be completely dissolved and the syrup cooled before it is fed to the bees. I do not add any medication, unless it has been found that disease is present in a colony.

The simplest feeders are the best. Twenty-eight pound lever-lid tins with tiny holes punched in the lid, or large pickle jars with screw-on lids (these can be obtained, usually free of charge, from cafes or chip shops) - are excellent feeders.

There is no advantage in feeding small quantities over a period of days or weeks. I prefer to feed colonies in massive doses and I like to get it all over before the end of September.

All hives should be water-tight, fitted with mouse-guards and securely fastened against winter-winds. Dampness is the worst enemy of over-wintering colonies. An excellent way of cutting down dampness is to have bottomless trays, about two-thirds of the depth of a super but otherwise, of the same outer dimensions as a super. The feedhole of the cover board should be covered by a piece of perforated zinc and the tray placed on top and filled up with wood shavings which will absorb moisture and help to conserve heat. In the working season they also provide a ready-to-hand supply of smoker fuel.

To sum up, my recipe for successful over-wintering of honey bee colonies is:

A colony strong in young bees.

Queen-right in every sense.

Adequate and ripe stores, about 50 lb.

Bees and brood free from disease or infestation.

The hive weather and mouse proof.

If the beekeeper faithfully discharges these simple duties he can, with clear conscience, send his bees into winter and expect them to emerge in the spring in good heart. All that will be required during the winter and early spring are periodic checks to make sure that all is well. If at any time it is found that the bees are right up to the food hole it could well be that the colony is running short of stores. This simply means that the beekeeper has failed to ensure that sufficient feeding was available in the autumn. The objective is to ensure that the colony has sufficient food to see it through to the first considerable intake of spring nectar. Winter feeding is really in the nature of first aid only, made necessary by the beekeeper being remiss in his duties the previous autumn.

As spring advances the bees may be stimulated by scarifying sealed stores with a fork or other pointed instrument. The bees will consume the exposed honey which contains pollen and it is the protein in the pollen that encourages brood rearing. It is sometimes recommended to feed bees weak sugar syrup in the spring to stimulate brood rearing, but I have found no merit in this practice. Only two things can stimulate brood rearing in the spring. One is water, but in these islands water is rarely a scarce commodity. The other is the provision of protein as mentioned above. Only if a colony is running short of stores will the feeding of sugar syrup in the spring be beneficial.

**5**

# Chapter Three

Early spring manipulation should be regulated by the weather rather than the calendar, or the impatience of the beekeeper. March is the danger month for colonies and it must be emphasised that constant check must be kept on stores and activity at the hive entrance. Colonies which have come through the winter may yet perish this month or in early April, brood rearing makes great demands upon stores.

Experienced beekeepers have an additional means of gauging the amount of stores in the hives - they 'heft' each hive by tipping it slightly backwards or sideways, lifting by the floorboard. This early in the year colonies must not be seriously disturbed. On a warm day towards the end of April, it may be safe to remove the crown board, or peel back the quilt - if quilts are in use. It will be found that the bees are occupying the upper of the boxes and a quick glance down between combs may reveal brood rearing in progress. In fact the beekeeper ought to know already whether brood rearing is taking place. First of all, if it is, the bees will have been carrying in pollen for some weeks and, secondly, a hand placed on top of the crown board will detect an increase in temperature. Normally the queen will be laying in the top box - this is one of the main advantages of having put the permanent super under the deep box in the autumn. It greatly facilitates the work of the beekeeper at this time of the year. Stores can be more fully verified and he can easily satisfy himself as to the queen-rightness of the colony. Brood should only be exposed briefly as the ambient temperature is not yet high enough for a more extended examination of the brood chamber. As spring advances and temperature rises it will become possible to spend more time examining the brood but temperature is at all times the deciding factor. This is a most important time because, very largely, successful swarm control depends upon good spring management. It will be found that the brood nest, that is, the combs actually containing brood, will be found towards the more sheltered side of the brood box and will be bounded on either side by a comb containing mainly pollen. Beekeepers sometimes speak of *"pollen clogged combs"*. In fact these combs are the colony's store of protein. The beekeeper's objective at this stage should be to do his best to ensure the progressive and natural expansion of the brood nest. To this end he will find it useful to move these pollen combs one space outwards by introducing one of the outer, empty combs between the pollen and the actual brood. This provides immediate additional laying space, if the colony is strong enough to make use of it, without running the risk of chilled brood. At the same time he will gradually move the entire brood over until it is centred in the brood box. The above operation does not involve brood spreading, which is an operation to be indulged in only by the experienced beekeeper. What can be done later when warm weather is more or less assured is to move one of the pollen combs right into the centre of the brood nest. This induces the bees to clear the comb of all pollen and in so doing they consume some of it, stimulating the production of chyle food and hence, more brood. Not only so, but it offers more egg laying space right in the heart of the brood nest.

As already stated it is in the competence of spring management that true swarm control lies. There must be no unnecessary disruption of the brood nest - all of the above can be carried out without pulling the brood combs apart. Robert Brown was wont to say *"If you go into the brood nest looking for swarm cells, you may not find them the first time, but you are likely to find them next time. You caused them!"*

In these islands, the first appreciable nectar flow comes in April from the plane or sycamore tree. True there are earlier, but minor sources of nectar, but these are significant only because they contribute to the general build up of colonies. The plane tree is not a dependable nectar source because the weather is so fickle in spring - in fact it is a characteristic of the British weather that it is fickle all the year round. But more about honey sources later.

The colony should now be building up rapidly. As soon as the top box is so filled with brood that the bees are covering the inner sides of the outer combs on either side, the two boxes should be reversed, bringing the bottom box back on top and the deep chamber containing brood should now be on the bottom board. There will be little, if any brood in the permanent super, but the queen will now move up into it, seeking the new egg laying space. If the flow is generous as may be hoped for, the result will be a crown of honey right across the top of the frames in the upper box. If this crown of honey is obtained, it is possible to dispense with a queen excluder for the queen will not, normally, cross sealed honey to lay beyond it. Sealed honey is the natural boundary of the brood nest. From the centre where the brood is, the order, outwards and upwards is - pollen, unsealed honey, sealed honey.

When the reversing operation is being carried out, opportunity can be taken to replace the floor board with a clean one. This is the only 'spring clean' that need be carried out. In fact the bees get on very well without any 'spring clean' by humans. I have seen beekeepers removing all 'brace' comb from the tops of frames. This is, of course, replaced by the bees and will be there to offend the beekeeper next time he examines his colonies - and a lot of honey will have been consumed by the bees to produce the necessary wax!

All the work in the apiary up to this point can be done without any serious disruption of the colony. Avoidance of disruption helps the build up of the colony and helps to keep the foraging force intact for the main honey flow.

Having said this, however, I have found it profitable to split the previous year's main honey producing colonies. This may appear to contradict the idea of keeping the foraging force of the colonies intact, but it is my experience that if a colony has been a main cropper the one year it can greatly benefit from being divided the following year. Each division seems to be rejuvenated as it is, indeed, by a natural swarm, properly handled. And these successful honey producing colonies are the ones from which to raise new queens.

As spring progresses it will become necessary to add additional supers. The beekeeper should guided as to when to do this by the same indicator as already given for the reversal of the brood boxes and permanent super, namely when the bees are occupying the inner faces of the outer comb on either side. Some textbooks recommend waiting until the bees begin to shoulder out the super combs. In my view this is too late. Such a colony is already congested.

I have already suggested that it is profitable to divide certain colonies. This is one of the few occasions when it may be necessary to actually see the queen. This is one of the operations which may call for good eyesight and a great deal of patience. The golden rule is to use as little smoke as possible, and never to puff the smoke down between frames. Smoke will drive the queen away from the area where she has been laying her eggs and she may go to any part of the after that. A gentle puff of smoke at the entrance and a gentle puff or two *across* the tops of the frames. It is advisable to use manipulating cloths, exposing only a few frames at a time. It is better to work with the sun behind you. In this way you can see right down to the bottom of the brood cells, illuminating eggs and larvae in all stages. Lift out each comb carefully and slowly and while doing this, scan the newly exposed side of the following comb. The queen may be seen on this comb side, scurrying away from the light. Usually her movements and her lighter coloured legs reveal her presence. Each comb must be carefully examined. Freshly laid eggs indicate that the queen has been at work on the comb - hence the reason for a restrained use of smoke. In fact this is a pointer as to whether or not the colony is contemplating swarming. If this is so the queen will have gone off egg-laying and will have deserted the outer brood combs which will contain only the older larvae or sealed brood. Mind you, nothing is certain in beekeeping! In a really powerful colony it is not always easy to locate the queen and if the brood nest is of more than one chamber, a separation of the parts for a few hours will make it so much easier. Slip a queen excluder between the boxes and a few hours later it will be possible to decide in which box the queen is to be found. Ideally, with a restrained use of smoke and a careful manipulation of the combs the queen may be found busily laying, surrounded by her retinue of workers.

A small colony should never be divided. During the build-up period a careful watch should be kept on the brood pattern. Brood should be in solid slabs with only a very light scatter of empty cells. These occur from several different causes. A few cells may still be occupied, or be used for storage when the queen reaches them in her egg laying and of course, when she is in full lay it would not be unusual, or a cause for concern, that a few of them proved infertile, provided these are few in number. The characteristic egg-laying pattern is a great arc stretching upwards and across the brood comb. It is normal for there to be a band of unsealed brood, or eggs showing up in this pattern. The queen makes regular courses of egg laying as brood hatches and cells are prepared for her to lay in.

# Chapter Four

I have long since got rid of gadgets in my beekeeping. If an appliance is not simple, practical and effective, it has no place in my beekeeping. But there seems to be a compulsion by beekeepers to 'invent' this or that gadget which will revolutionise beekeeping! Worse still, beekeepers want to make their own hives usually with their own 'improvements'. Usually these finish up as kindling wood. However the double screen board to which I was introduced many years ago by a Canadian beekeeper was no useless gadget. I shall be forever grateful to him. I can, therefore, claim no credit for this appliance. But it is very practical indeed. It is in use in many parts of the United States and Canada. I flatter myself that I have persuaded a fair number of beekeepers in this country of its merits. It is certainly one of the most useful and versatile of appliances in the apiary.

It is simply a crown board from which the centre area of wood has been replaced by two sheets of wire gauze, or perforated zinc, one above the other and three-eighths of an inch, or a bee space apart. Not less than one-third of the central area of the crown board should be constructed in this way. On one side a piece of the fillet of wood which makes the bee space on the crown board is cut away at an angle and is fixed in place with a central screw so that it can be pivoted open or shut to make or to close a side or rear entrance. This works rather like the old-fashioned way of making a simple 'button' for a wooden door or lid. The whole idea is to keep separate the bees above and below the screen board. With the two layers of wire gauze between the two lots of bees, colony odour is permitted to circulate between the two lots of bees, but the queens cannot sting one another. This double screen board lends itself to many major manipulations of the honey bee colony as will be seen in the ensuing pages.

# Canadian Screen Board

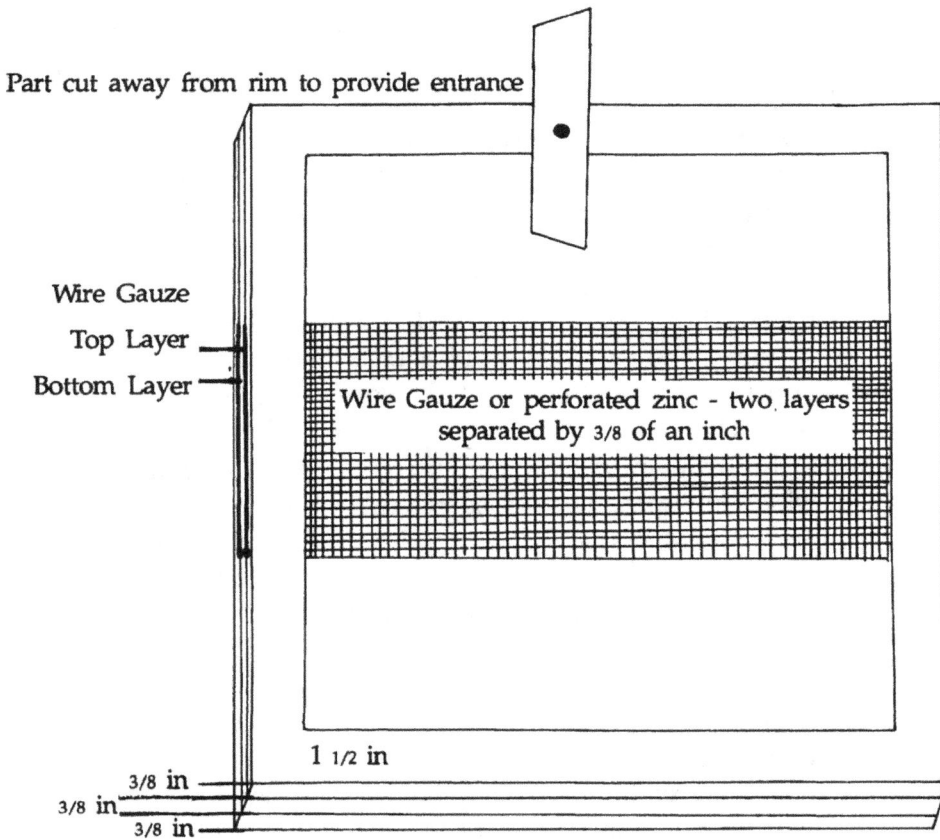

Part cut away from rim to provide entrance

Wire Gauze

Top Layer

Bottom Layer

Wire Gauze or perforated zinc - two layers separated by 3/8 of an inch

1 1/2 in

3/8 in

3/8 in

3/8 in

## Dividing a colony which is not raising queen cells.

The first step is to select your breeder colony, that is a colony whose bees possess the best combination of desirable qualities from a honey producing point of view, for queen should be raised only from the best available stock. But more of that later.

A deep frame should be fitted with shallow super foundation, preferably unwired. This foundation should be scalloped by removing triangular pieces from the bottom edge. The frame is now inserted into the middle of the breeder colony, that is, the colony from which you wish to raise queens. This is done two days before the projected colony division.

Now to deal with the colony which is to be divided. As far as possible the combs containing sealed brood only should be separated from combs containing eggs or larvae. The scalloped frame should now be removed from the breeder colony and the adhering bees gently brushed from it. It ought now to be well laid up with eggs. The beekeeper must decide how many queens he wishes to raise. The required number of cells containing eggs are selected on the scalloped edges of the comb. The cells surrounding each selected cells should have their eggs destroyed. This is to allow the resultant cells to be cut out without damaging the queen cells. This frame should be carefully placed in the middle of the box containing the sealed brood and this box is placed on the bottom board and it should have its quota of frames made up with frames of drawn comb. A queen excluder is now placed upon this box. If there are supers containing honey, but no brood, these are placed over the queen excluder. This is the nursery colony. In the event that there are no supers, the queen excluder can be dispensed with and the double screen board is placed on top of the colony with an entrance open to one side or to the rear of the hive. This entrance must be on the top side of the screen board for it is to serve the colony which is placed over it. This colony is made up of the unsealed brood and eggs together with the old queen. The hive is then closed up, with the crown board and roof in position.

The flying bees from this top colony will all join the bottom lot with the queen cells now being built. This makes the bottom part the more prosperous part of the divided colony. Queens ought to be reared 'only' in prosperous conditions. Three days later the bottom brood box must be thoroughly examined and all un-wanted queen cells removed. This is a simple method of raising about as many queens as the average beekeeper is likely to require in a season. By this method the beekeeper knows exactly when the young queens will emerge. The ripe queen cells can be easily cut out and deployed as the beekeeper may decide.

Meantime in the parent colony above the screen board the old queen will have benefited from a rest from laying in a very large colony and the colony will have benefited from the warmth rising from the bottom colony. The two lots can be kept *in situ* until the onset of the first major honey flow. The top colony can be relocated on a new stance, creating an additional colony.

If colony increase is not required, the old queen can be removed and the two lots of bees brought together. This is particularly easy when a nectar flow is in progress. The bees will unite without trouble. The screen-board is turned upside down with the open entrance retained on the same side of the hive as before. It is placed over the united colony with an empty brood chamber, or super, placed on top to lift the lid clear of this top entrance. The bees will continue to use this entrance for a few days but they quickly find that they are joining the main colony below the screen-board. This use of the top entrance will gradu-ally cease and the main foraging strength of the colony has been kept together and the risk of unwanted swarming has been greatly reduced.

When I was first introduced to the screen-board by my Canadian friend he said to me *"I would dearly like to show Scottish beekeepers what a good double-queen colony could do with all that glorious ling heather."* Well, I have followed his advice over a number of years and have found that, given the right conditions, these colonies fully justify his ambitions. Unfortunately, no matter what we do in beekeeping, our British weather always holds the trump card! Undoubtedly some spectacular takes of ling honey can be obtained by these double queen colonies which always abound in large numbers of field bees, and be still strong in young bees, they also winter very well indeed.

## Dividing a Colony with Queen Cells.

If it is found that a colony is actively preparing to swarm and queen cells are present - and occupied, the screen-board greatly simplifies what has to be done. The queen must be found and this time she is given mainly hatching brood or older larvae. All queen cells must be removed from the combs which accompany the queen. The queen and her comb are carefully set aside. The remainder of the brood nest is thoroughly searched and all queen cells except one are removed. I invariably choose an open but occupied cell - I then know that I have a viable larva and have a good idea of the time the young queen will hatch. The comb with the queen cell is marked with a thumb tack pushed into the top bar. I say that advisedly, because I once found that a beekeeper had pushed the tack into the queen cell!

Both brood boxes should now be filled up by inserting the required number of frames of drawn comb. The brood box with the selected queen cell is placed on the bottom board, together with any honey supers  no brood in the supers! The screen-board goes on top as described already. The queen with her brood and bees are placed over the screen-board. The flying bees from this brood chamber will rejoin the bottom colony. The top lot cannot swarm because it has lost all of its flying bees; the bottom lot cannot swarm because it does not have a running queen - always assuming that the beekeeper has done his work well and has not left an additional queen cell. Very often the reaction of the beekeeper on seeing queen cells is to cut them out. In fact this is just about some beekeepers only attempt at swarm control. Swarm cells are the best queen cells that a colony is capable of producing and they ought not to be destroyed without careful thought. Any subsequent cells are likely to be inferior. Once again the beekeeper can choose what to do with the additional colony that has been formed with the old queen.

# Chapter Five

No matter what method of swarm control may be adopted, swarms will still occur. For a long period of time beekeepers were advised that swarms were evidence of poor beekeeping practice and that swarming was an unmitigated disaster from a honey production point of view. Not so. What is important is how both swarm and parent colony are dealt with. The first thing is to establish which is the parent colony. This is not always easy as several colonies may have reached swarming point together. The observant beekeeper may well have noticed tell tale signs of swarming and will make every effort to circumvent it. Only experience can really guide the beekeeper in all of this. Two much favoured ways of identifying swarming colonies are:

(a) Clipping the queen's wing to prevent her absconding with a prime swarm. This procedure certainly gives the beekeeper a second chance to deal with the colony. Queens must be handled delicately but surely and only one wing should be clipped using very fine scissors. This operation should be carried out inside a hut or a car lest the queen fly off and be lost. It is useful to practice handling queens by using drones which cannot sting.

The problem with clipped queens is that a swarm may still come off and the clipped queen will succeed in flying a short distance before falling to the ground, often amongst long grass where she may be overlooked or stepped upon. Unless the beekeeper is on hand the swarm will eventually return to the hive - minus the queen, only to come off again with the first virgin queen to hatch and may fly a great distance and be lost to the beekeeper.

(b) If the queen is marked by putting a small dot of colour on her thorax (after she has been mated and is laying eggs) then her absence from the hive and the identity of the parent colony will be much easier to establish. It used to be that each queen was marked with a different colour so that would be immediately apparent from which colony the swarm had emerged, but nowadays the recommendation that all queens be marked with a colour denoting their year of birth, rather than its particular colony, makes the beekeeper's task much more difficult. The old recommendation that some bees from the swarm be dusted with flour and tossed into the air so that they will return to their hive is utterly useless for it will be found that bees from a swarm will go into different hives. The only real guide is to examine colonies and if one is seriously depleted in bees, that is the one likely to be the swarmed colony. There will normally be no eggs in the brood nest as the queen will have gone off the lay in preparation for her flight.

Once it has been established which colony is the parent one the work of dealing with the swarm can begin. The brood nest with all its brood should be set aside and a new brood chamber, fitted with frames of foundation put in its place. The swarm can either be dumped in from the top or the beekeeper can indulge himself by running it at the entrance - there is no thrill quite like it in beekeeping! A queen excluder should be placed over the swarm and any honey supers added - again these must be free of brood. The screen-board is placed in position on top and the parent brood chamber(s), with a selected queen cell is placed above the screen-board with a side or rear entrance open. All the flying bees from the top colony will rejoin the swarm below, keeping the foraging force of the colony intact. Neither section will readily swarm again. The bottom lot has a lot of comb building to do and the young queen has to hatch, be mated and begin laying while the top lot has lost all flying bees. By this method the colony is requeened, the swarming urge has been satisfied. Both queens may be permitted to go on laying for as long as the beekeeper decides. These colonies united as already described and headed by the young queen will give good account of themselves during the main honey flow and will be ideal for heather going.

## Making Nuclei.

Any one of the above procedures will lend itself to the making of nuclei, because each produces a number of queen cells. Nuclei ought to be made as early in the season as is practicable and should not be of less than four or five frames, three of which

**16**

should be brood in all stages. From the time a nucleus is made until a young queen is hatched, mated, laying and her brood emerging is something of the order of five weeks. During this time a nucleus is a wasting asset, unless it is well furnished with hatching brood. Nuclei should be available in the apiary right throughout the summer honey flow in case one of your honey producing colonies becomes queenless at a critical time. In the event that your nuclei are not required for requeening they will build up for the ling heather flow. Every unit in the apiary must pay its way. A good ratio is one nucleus to every four honey producing colonies.

## Over Wintering Nuclei

The double screen board comes in useful for over-wintering nuclei or weaker colonies. The only drawback - if it is one - is that the beekeeper must ensure that each nucleus is furnished with adequate stores to see it safely through the winter and well into the spring. In actual fact, however, there are feeders available which are attached to the outside of the hive box but which conduct the feed into the hive where it is accessible to the bees. Using screen-boards to separate them, nuclei can be stacked on top of each other for the winter months, with entrances on different sides of the hives. Each benefits from the warmth generated by the whole. Nuclei over-wintered in this fashion emerge in the spring in excellent condition.

Each must be removed to a permanent stand before the foragers begin flying freely.

# Chapter Six

Many theories are postulated to explain the swarming of honey bees and countless systems have been devised to eliminate swarming. I cannot claim to have accomplished either, the best I have been able to do is to reduce the incidence of swarming. The plain truth is that swarming is a basic necessity of honey bee life. Bees reproduce on two levels. They reproduce individual bees - eggs, larvae, bees, and they also reproduce colonies in order to replace those that perish for one reason or other. A non-swarming strain of bee is not only unobtainable, it would also be undesirable.

What is desirable is a strain of bee which is not given to excessive swarming. I can remember, in my early beekeeping days, purchasing a six frame colony of Dutch bees which had been imported to pollinate apple trees in the south of England (it cost me one pound and ten shillings, or 150p).

This colony built up rapidly - but it also built more than twenty queen cells and it was ready to swarm even although it had plenty of room for all purposes. Swarming is often attributed to congestion in the hive. Undoubtedly congestion is an important factor. But congestion takes many forms, all of which may set off the swarming impulse. If the queen is severely restricted in her egg laying, that is a form of congestion which is related to swarming. Probably the bees feel that the queen is failing and takes steps to supersede her which then turns into swarming. The size of the brood nest should be measured by the number of cells available to the queen for egg laying. It has been said that a swarm is more in the nature of a funeral rather than a wedding, for it means that the queen may be failing in her laying. There is also congestion if the nectar gatherers run out of room to store their freshly gathered nectar - storage space must be given in advance of the immediate need for it.

So many textbooks advise that the time to super a colony is when the bees shoulder out the tops of the combs to accommodate incoming nectar. Too late - that colony is already congested. A new super should be given when the bees are occupying the inner faces of the outer combs on either side of the hive.

Every strong colony should be given a deep chamber of foundation in order that the bees at the right physiological stage for producing wax are given the opportunity to build comb, otherwise there may be another form of congestion in the hive. This deep box should always be placed *above* the brood nest. Only in this way will really good brood comb be obtained - the bees will build worker comb right down to the bottom. Possibly these combs will be filled with honey, if a good flow is in progress. After this honey has been extracted, the beekeeper will have a box of first class brood combs. There are only two ways to get deep combs this good:

(a) by the method just described or

(b) by hiving a prime swarm on foundation.

Introducing a queen excluder below a deep or shallow of foundation can often produce congestion in the brood nest for the bees do not readily take nectar through a queen excluder and may very well squeeze the queen out of brood combs by storing nectar or ripe honey below the queen excluder. The beekeeper should aim to reduce swarming by providing for the expansion of all hive activities in advance.

Weather conditions can be a contributing factor to swarming. If colonies are at their peak of population and they are confined by inclement weather as happens so often in these islands, the first sunshine can often bring out a swarm, for colonies often raise queen cells in these circumstances.

*"Satan still some mischief finds for idle hands or bees to do."*

The queen gradually extends her egg laying to reach her peak of around 1800 eggs per day around the beginning of June. She will maintain her peak for several days, and then her egg laying will gradually decline right into late autumn - in fact brood rearing continues intermittently, and at a very low level throughout the winter months.

Brood food is produced by young bees - it is their very first task in life to feed the larvae. As egg laying declines a point is reached when the number of young nurse bees greatly exceeds the number of larvae to be fed. In most cases the excess brood food is consumed by the bees themselves, but given the right conditions, e.g. bad weather or, more particularly, unemployed wax producers, brood food is deposited in large quantities in a

**19**

few cells. Other factors of congestion may coincide with an excess of brood food and swarming is likely to result.

## Collecting a Swarm

A useful piece of equipment for collecting swarms is a six-frame nucleus box which is about four inches deeper than the British Standard Frame. A swarm settles down quickly if the box contains at least one drawn comb. This swarm box should be so constructed that a temporary crown board most of whose area is of perforated zinc or wire gauze and which can be fixed in position so that the bees cannot escape while they are being transported. This crown board should also be fitted with a good carrying handle.

Not all swarms settle in convenient places and it may be difficult to take the whole cluster in one go. However, even if the queen is not taken with the body of bees which has been taken in the box or skep, everything is much easier if the queen is taken with the greater part of the cluster - if the bees actually taken are calling, and if you keep disturbing the cluster remaining where the swarm had alighted, preventing them from settling there, they will gradually drift down to join the bees in the box and the queen will also join them there.

Smoke should be directed above the cluster, preventing it from climbing higher and out of reach. The cluster must be constantly disturbed. Rather a lengthy process, but it works. Fortunately not all swarms alight in inaccessible places. From time to time claims are made about the elimination of swarming or the discovery of some particular key to the problem, but these must be discounted. It is wise to remember the words of that great American authority on beekeeping, the late Dr. Miller:

*"Of all my work in beekeeping I have had least success in controlling swarming."*

The only person who has sought to put the making of an artificial swarm on a scientific basis is Professor Taranov of the U.S.S.R. He published his method, using his swarm board, some thirty years ago. The Taranov method is reproduced with acknowledgment to the 'Bee World' 1952.

**20**

# TARANOV'S SWARM BOARD
"Swarming Bees and a Method of Segregating them."

*In the Spring 27 to 35 bees are needed to cover 100 brood cells. A the bees emerge they remain on the comb, forcing other bees off. These occupy the combs next to the brood and are ready to attend to the queen and feed her.*

*Field bees pay no attention to the queen, and she does not lay eggs on combs devoid of nurse bees.*

*With the approach of summer and of higher temperatures less bees are needed to hatch the brood but their numbers continue to increase until it is far in excess of the needs of the colony.The nurse bees then begin to crowd round the queen continually offering her food. This at first increases the laying of the queen, but soon she is unable to cope with all the food offered her, and refuses it.*

*The bees then come quite close to the queen and occasionally a bee jumps on her back and shakes herself for a few seconds. Finally the bees start queen cells and literally chase the queen until she lays eggs in them. When these eggs are laid the bees stop feeding her. They do not work but hang about the hive in festoons.*

*It was obvious that these were the swarming bees, and an additional proof was provided by counting the foragers before and after the issue of the swarm. Their numbers tallied. These idle bees are up to 21 days old, but physiologically they are all young, because they have not been able to perform any of the duties appropriate to their age.*

*The problem then arose how to separate them from the rest of the bees. Taranov used a special kind of board. It consists of two boards 20 inches long and the width of the alighting board. Two ends are nailed together and at the other end two supports, the height of the alighting board, are placed below them. This provides a ramp sloping up from ground level to the level of the alighting board.*

*A cloth about 36 inches long is spread in front of the colony to be dealt with, and the board placed upon it, so that the higher end is 4 inches away from the alighting board. Remove the stock to one side an put a new hive in its place. Every brood comb is then put into the new hive, after all the bees have been shaken or brushed on to the cloth. The same procedure is then applied to the supers if any.*

*The bees move up the board towards the hive and when they reach the top, the swarming bees go under the board and the queen joins them.*

The other bees fly across the 4 inch. gap on to the alighting board and enter the hive.

In about 1 1/2 hours the bees are settled, those under the board forming a cluster.

To check these experiments, Taranov weighed the swarm so obtained and reunited it with its parent colony. The next day he swarmed them again, and the weight of the cluster was the same. He repeated this at intervals and found that the weight gradually increased as the time of the natural swarm approached

These artificial swarms had all the characteristics of a natural swarm:

1. If put into another hive with the queen they remained  there, even if next to the parent hive.

2. If hived without the queen they returned to the parent  hive, except for the very young bees who were too young  to have made orientation flights.

3. They worked energetically.

The only difference between these artificial swarms and a natural swarm was the presence of very young bees, and the absence of bees from other hives.

The advantages are obvious. It eliminates the actual issue of a swarm and their clustering in an inaccessible place. It can be done at a convenient time, and finally, it cuts short the period of idleness of the young bees.

Another advantage is that once the board has been used, and the parent colony dealt with in the usual way, no more swarming occurs, and the detailed examination for queen cells is unnecessary, which lightens the work.

It is important that the hive put in place of the parent stocks should be of the same type, otherwise the foragers are reluctant to enter.

M. Simpson in "The Bee World" 1952


I have already described how to deal with swarm and parent colony using the double screen-board. However, if you do not use the screen-board it is still good practice to move the parent brood nest off its stand and at right angles to its original position.

A brood chamber containing a full complement of frames fitted with foundation is placed on the parent stance. If you have hived the swarm on foundation - with one drawn comb, it is easy to lift these into the new brood chamber and make it up with further frames. A queen excluder should be placed over this chamber and any honey supers added above the queen excluder, add the crown board and roof. It is wise to go through the parent brood chamber removing all unwanted queen cells, or splitting it up into nuclei. This colony, if left in this position will be stripped of its flying bees which will return to the swarm. A swarm and its parent colony, properly handled will give very good account of itself.

Swarming is not a sign of poor beekeeping and even the loss of a swarm is not a complete disaster. The main worry about the loss of a swarm is that it may cause some upset to the uninitiated who may find this great cluster of bees in their garden, or elsewhere in their property, but so far as the beekeeper is concerned it merely represents the temporary weakening of a colony. Nature is a wonderful provider and a colony so weakened will quickly make up its loss of foraging bees and will be rejuvenated by a young queen raised under the very best of conditions - better far than many produced under the manipulation of the beekeeper.

# Chapter Seven

Strain of bee is of great importance in successful honey production. It is not very meaningful to talk of the various races of bee because the honey-bee population of the British Isles - like the human population - is an amalgam of many races.The beekeeper must, therefore, try to work towards a strain of bee which best suits his purpose. What are the desirable characteristics for which the beekeeper will look? They are:

(a) reasonable temper

(b) good working strain

(c) not given to excessive swarming

(d) sealing the honey combs with good clear cappings

There is no real pleasure in keeping bees which attack whenever one comes near the apiary and which will pursue after one leaves the apiary. Bees should, given proper handling, remain relatively quiet on the comb and not be rushing about. A great deal has been said about bees working early in the morning and late in the evening. I have never been convinced that this was an important characteristic of bees: bees work when plants are yielding nectar and that is very largely controlled by temperature and other weather conditions, but I have included this in my list of desirable characteristics.

Undoubtedly there are strains of bees which build swarm cells even when there is no apparent reason for so doing - and they usually build a greater number of cells than colonies less eager to forsake their hives. Some bees seal their honey combs without an air space between the cappings and the honey and this results in greasy, wet looking honey combs which are not very attractive to look at and make uncapping more messy than it ought to be. It is unfortunate that no real attempt has been made in this country to improve the strain of honey bees available to British beekeepers.

The honourable exception to this must be the activity of the British Isles Bee Breeders' Association. But with the limited resources at its disposal it has not really accomplished much of significance. It was a long cherished hope of mine that a queen breeding station, working in a controlled breeding conditions, would be established in my beloved Scotland and serving all Scottish beekeepers, but I have failed completely to persuade my fellow beekeepers that this is a truly practical project. *"It would never pay." "Who would put up the money?" "We have too few beekeepers." "Our climate would not make it possible."* I have heard these incantations time without number. Yet Norway, Sweden, Denmark, Finland - and other small countries, with even more inclement weather than ours have established successful queen breeding stations.

Short of establishing such a facility beekeepers can contribute to the general beekeeping weal by taking care to keep reliable records of the performance of each colony. Queens should be raised from colonies which have the best combination of the qualities already outlined - even if they are not all fully achieved. Every bit as much care should be exercised to ensure that as many drones as possible, of similar high quality are available when the young queens fly forth on their nuptial flights.As an Ayrshire man I am well aware that the bull is every bit as important as the cow in  maintaining herds of high milk yield. All this may seem to be the counsel of unattainable perfection. Perfection in any field is difficult, or impossible to attain, but it should always be the golden target at which to aim.  Just maybe events have moved in such a way as to make queen breeding on a commercial scale possible. The first thing that has happened is that with the ban on importing bees and queens from overseas, the price of a queen has risen steeply and, secondly, the three agricultural colleges of Scotland - and their English counterparts - have been placed on a commercial basis, required to make a profit. I commend to them the idea that they could now make a tidy profit by raising and selling high quality honeybee queens, but they must be bred selectively.

# Queen Introduction

I have already spoken of the importance of nuclei. If the greatest benefit is to be derived from 'nucs' they must be made as early as is practicable - late May or early June in South-West Scotland - and they must be made sufficiently strong in bees and brood to earn their keep in the current year. I have already given my recipe for the making of a good nucleus. In a well run apiary there will be a minimum of one nucleus for every four honey producing stocks. If such a colony needs to be requeened, by far the safest method is by uniting a queen-right nucleus to it. I have never known this to be unsuccessful.

Colonies which have been queenless for a long time need to be conditioned to receive a queen. First of all they should be fed heavily and secondly they should be given at least one frame of hatching brood (without adhering bees) a few days before requeening is attempted. Feeding creates conditions of prosperity which facilitates just about any manipulation of a colony. The hatching brood will provide a force of young bees at the right physiological stage to care for the larvae which will soon appear once the queen begins to lay. There are many ways of introducing a queen.

Brother Adam has proved that it is possible to persuade a colony to accept a queen if the old queen is removed from the comb on which she has been laying and the new queen is placed in exactly the same place on the comb, immediately. Most beekeepers use one or other of the queen introduction cages readily available from dealers. It is absolutely essential in all attempts at requeening to make sure that the colony is, in fact, queenless. A fairly obvious instruction, you may say - but it is astonishing how often beekeepers fail to make sure. My favourite method of queen introduction, apart from the queen-right nucleus is the paper bag method given on the next page.

## Paperbag Method of Queen Introduction

This is really an adaptation of the newspaper methods of uniting stocks to queen introduction. It was first published in the 'New Zealand Beekeeper', which is a most excellent journal, packed with practical beekeeping. *"Prepare a small paper bag such as is used for confectionery, by piercing it many times with a pin or other sharp instrument. The perforations should be in the lower half of the bag. From the hive which is to be requeened, remove a brood frame and brush about a tablespoonful of bees from the frame into the opened bag. Walking away from the apiary, shake the bag for one minute, by the watch.*

*It is now advisable to take the bag with bees and the queen either inside the house, or even a car. Let the queen run on the window pane, pick her up, open the bag - the bees will be lying in the bottom of the bag. Drop the queen, carefully, down amongst the bees, and making sure that she has not climbed up inside the bag, twist the neck of the bag to close it and tie it with a piece of thread. Make a space in the middle of the brood nest, drop the bag between frames and fix the thread to a frame with a drawing pin".*

I have never known this method to fail. In a matter of an hour or so little scraps of paper will be carried out of the hive. I always leave the colony undisturbed for at least three days.

Undoubtedly a factor in the success of this method is that the queen is accompanied by bees of the colony to which she is introduced. It has also been found that when a space is opened up between combs in the brood nest, young bees tend to cluster there and they will accept a queen more readily than older bees. I always feed such a colony immediately before - and immediately after carrying out this manipulation.

This is probably the most suitable juncture to mention laying workers. If a colony has been queenless for some considerable time some workers will actually manage to lay eggs. These can only hatch out as drones. Typical of the work of laying workers is that the brood area has no real pattern, eggs are laid randomly. Various manipulations have been advanced to get rid of laying workers but the one that really works, in my experience, is to introduce a virgin queen. Such a queen will make short work of the laying workers.

**27**

# Chapter 8

## Scottish Honey Sources

It may be argued that Scottish honey is little different from honey gathered elsewhere in these islands. Not quite true! The famous Loch Lomond of *"You tak' the high road and I'll tak' the low road"* fame marks an important area of change of plant life. It is the southern most area where northern types of plants grow naturally and it is the northern most area where southern type plants grow.

This mixture and separation of flora must produce distinctive types of honey. At the same time there are a number of nectar yielding plants common throughout these islands, but these, too, are influenced by such factors as local soil and weather, producing honey with slight nuances of flavour and colour.

A major factor in the climate of the West of Scotland is the benign influence of the Gulf Stream which washes right up the west coast completely changing our climate from what it would otherwise be. It greatly modifies temperature and brings so much more rain, which is the beekeeper's greatest problem in Scotland. Dampness kills off more colonies than any disease.

Our prevailing winds are south-westerly and our mountains cause these winds to deposit the moisture they have absorbed over some three thousand miles of the Atlantic.

Ignoring minor sources our first appreciable nectar flow is from the plane or sycamore tree. This yields a delicious dark greenish-brown honey which granulates to a smooth white honey. *"Dark honey for pale people"* is a good selling point. This is not a very reliable source as the rhesoms or florets, do not yield readily unless the weather is reasonably warm and dry and in a Scottish April that does not happen too often. In favoured areas the plane tree flow will be followed by a considerable flow from the soft fruits and the apples, pears and plums. The Clyde valley was a prime fruit growing area, but alas, the fruit industry no longer prospers in the Clyde valley. The Carse of Gowrie remains a particularly good fruit growing area with many acres of raspberry which induces beekeepers for miles around to move their bees in to take advantage of the raspberry crop.

A more recent source of honey is now available in many parts of Scotland. Farmers are growing oil seed rape on an ever increasing scale. Undoubtedly it is one of the most generous of nectar yielding plants. Its bright yellow blossoms prove irresistible to bees who will forsake all other foraging to work on this source of nectar, attracted by its sweet scent. It has been estimated that as much as 500 lb of honey may be collected from a single acre of rape. There are two main types of oil seed rape,

(a) Winter rape, commonly called 'Swede Rape'. This is sown in the autumn and blossoms April-May of the following year. This is the most profuse yielder of nectar:

(b) Summer rape or 'Turnip Rape' which is planted in the spring but yields at about the same time as the 'Swede' rape or in June or July, depending upon climatic conditions, filling in the notorious 'June Gap' in the beekeeping calendar when little else yields. Oil seed rape is a member of the brassica family.

It is not an unmixed blessing. Rape honey granulates with astonishing rapidity. Supers are filled rapidly and the beekeeper must not wait until all of the honey is sealed, as with other honey. When they are well filled the combs should be taken out and given a shake. If the honey remains in the comb - even if it is not fully sealed the unsealed honey should be extracted and sieved through a very coarse cloth. Fine muslin will not let rape honey through. It is recommended to warm the honey all the way through until it is finally put in the jars. If it is put into 28 lb cans or into buckets and is stirred vigorously and frequently it will cream and it can then be put into jars. This unsealed honey should be consumed within a matter of weeks. It should not be exposed to air more than is absolutely necessary for it is highly hygroscopic - that is, it absorbs moisture from the atmosphere, as indeed does all honey but none so rapidly as rape honey. If these precautions are neglected it will ferment in no time.

Meanwhile the sealed honey should be uncapped, extracted and put in jars separately and without delay. Otherwise the honey will set rock hard in the combs in a matter of days, or even hours - and not even the bees will be able to use it. If this honey is not processed without delay the only remedy will be to melt down comb and honey together.

Oil seed rape granulates with an exceedingly coarse grain unless it is stirred thoroughly and frequently. Extracted combs should be given back to the bees immediately to be cleaned up - but even this can lead to granulation problems within the hive and with subsequent honey crops.

Beekeepers are now adopting a different method of dealing with oil seed rape honey. It is as follows: The super combs are fitted with 'starter' pieces of foundation, that is strips of unwired foundation of about half-inch depth. The bees quickly complete building the comb and filling it with the ungathered honey. This honey should be left until the combs are completely sealed. Of course, it granulates. Supers are removed in the usual way and the honey is cut out of the frames. The combs are then chopped into small pieces and honey and wax are liquefied being careful that the heat does not exceed 120 F. (49 C.) it is then left to cool, the wax rises to the surface, together with any impurities. Once the wax has solidified it is removed and can be traded for foundation - any appliance dealer will accept wax on this basis. The honey is put into jars and because it was allowed to ripen fully there should be no trouble with fermentation. It is still a very messy process.

Oil seed rape honey has a light, golden colour, but has little flavour. The best that can be said for it is that it is pleasantly sweet. It needs to be blended with other honey which can impart flavour. No doubt, beekeepers will in time learn how to deal with the granulation problem. They will need to, for the bees will assiduously work any rape within flying distance, spurning any other source available at the time.

Since writing the above there is some indication that oil seed rape may be falling out of favour with Scottish farmers and that large areas of beans are being planted instead.

Recent medical research suggests that oil seed rape may cause certain types of cancer in humans, when it is in flower. It is thought that the pollen of this plant is the causative agent.

Over many years I have carefully watched the hawthorn blossom, but have rarely seen bees working it in any considerable numbers, even when it is profuse and weather is favourable. However, I accept the word of some beekeepers that they have had a crop from the hawthorn. I am told it has an almond flavour.

A few years ago I obtained a sample of '**hawthorn**' honey from a fellow beekeeper. I did the normal pollen analysis under the microscope. I did not identify any hawthorn pollen grains in the honey.

In some areas of Scotland a considerable crop can be obtained from the bell heather. This heather yields a beautiful '**port wine**' coloured honey with a delicious flavour, which can be extracted in the usual way. Bell heather honey, for some strange reason, is a much under-rated honey. It comes on flow mainly in the month of June into July. Sometimes it breaks into the '**June Gap**' - about fourteen days, when the bees find little other fodder,. Fortunate are the beekeepers who have bees within working distance of bell heather!  Towards the end of June and into July comes what for many areas is the main crop, namely white clover. Clover needs really warm, but not necessarily sunny days. Too hot, dry days, dry out the clover nectar. It is impossible in Scotland to get absolutely pure clover honey - or indeed any honey pure to one source, but predominately clover honey is a very light, sparkling honey, pale yellow and of a rather insipid taste.

Following close on the clover flow - end of July - beginning of August in South-West Scotland, comes the lime flow. Not all lime trees yield nectar, or yield every year. The nectar of certain types of lime tree seem to have a toxic effect on bees and the affected bees may be found staggering around on the ground beneath the trees. It also affects bumble bees. I am fortunate in that within easy foraging distance of my home apiary there are well over one thousand lime trees. In my beekeeping lime is so important that I am frequently torn between taking stocks to the ling heather or leaving them to get on with the lime flow. Lime honey is a sparkling pale green honey with a slight tart flavour. Mixed with the less characteristic clover honey it makes a delightful table honey.

There are other lesser sources which together contribute a fair amount to the annual honey crop. Early in my beekeeping, one year my bees collected a pale pink honey and the only explanation I could find for this honey was a profusion of red campion in a nearby stone quarry. Another year my bees worked a considerable area of onion which was being grown for seed. The hives reeked of onion for about ten days and I thought that honey and combs would be ruined by this rank odour. Quite quickly the smell dissipated and there was no taste of onion in the resultant honey.

Not far from my present location there is quite a lot of balsam along a river bank and the bees work it enthusiastically. Another source of no mean order is the fireweed, or willow herb - those tall spikes of pink flowers which seem to spring up in so many places. The honey from this plant is nearly water-white and again it is a honey with very little flavour or aroma.

Whatever the source of the honey, we in these islands are favoured with honey of the highest quality and flavours. It deserves to be prepared and packaged under the best possible conditions. Too often have I seen honey being extracted, strained and bottled under something less than hygienic conditions. We have a quality product, let us keep it that way right up until it is on the consumer's table! The extractor must be spotlessly clean - if you have a friendly dairyman you may persuade him to steam clean the extractor and honey tanks. The domestic dish washing machine makes an excellent job of cleaning the honey jars, leaving them gleaming. In between the extractor and the honey jars are the uncapping knife, the metal sieves and the filter cloth and - above all - the beekeeper's hands.

# Chapter Nine

Colonies for the heather should be carefully selected and adequately prepared. It is no good taking a colony which is headed by a queen which has already done a full season's work. The queen of a honeybee colony has been likened to the sparking plug of a car. It is the source of colony power. It is therefore essential that each colony be headed by a fresh young queen in full lay. Such a queen gives the whole colony impetus and ensures that there will be plenty of young bees to survive the heavy wear and tear of working the ling heather.

At the same time she should be confined to a single brood box so that the brood nest is compact. Under this brood box it is useful to put a super of drawn comb with some stores in it. It need not have a lot of stores, but it is a useful supplement to the food in the main brood chamber - bees can go short of food at the heather if the weather is unkind, as it often is. The main purpose of the bottom super is, however, to lift the colony away from the chill August night breezes.

In reducing the brood nest to a single deep there may be some surplus frames of brood which can be given to other, weaker colonies.

A screen board with at least three quarters of its area being of wire screen, or perforated zinc should be placed over the top box of colonies destined for the heather hills. These should be fixed in place by four screw nails. Entrances should be fully closed. It is useful to attach a piece of string to entrance blocks so that they can be removed to liberate the bees once the hives are on their heather stances.

There have been many devices for fixing the hive parts securely together while being transported to and from the heather. My favourite is the little metal buckle supplied by E. H. Thorne (Beehives), Wragby, Lincoln.

Using these buckles with a suitable length of plastic binding tape and provided the frames inside the hive are firmly fixed, e.g. by using Hoffman self-spacing frames, the hives could be turned over and over without crushing bees. Thorne's are now marketing a new kind of buckle which does not involve cutting the tape to release it, as the little buckles do. This new buckle is much more expensive and it does not appeal to me as does the little buckle. I prefer to use two buckles and tape to each hive but some of my fellow beekeepers tell me that one is quite sufficient.

Hive stands for the heather can be easily made by using two pieces of wood each half checked in the middle to make an 'X' or St. Andrew's cross.

It has been my practice to prepare the hives a day or two in advance of moving them to the heather, leaving only the closing up of them until late in the evening before departure. I found it best to move the bees early in the morning, normally leaving about 5.30 a.m. Roofs are removed and the hives, roofs, stands, shovel and the usual equipment for handling bees are loaded on to the lorry. Another useful piece of equipment is a ball of plasticine - to stop up any cracks or holes from which the bees may escape in transit. On arrival at the heather site places are levelled for each hive - hence the shovel. The hives should be set out in an irregular pattern to reduce the drifting of the bees. They should also be situated well away from other bees which have been moved to the heather, in order to reduce the possibility of infection from disease acarine is notorious for spreading in these conditions. Once the hives are in place roofs are put on, temporarily, and entrances are opened.

Now is the time to sit down to a well earned cup of tea, allowing the bees to settle after their journey. Soon the smoker will be lit, veils will be donned. The screen boards are removed and queen excluders are placed over the combs, then the supers and crown boards. Once the roofs are replaced the hives should either be tied or weighted down because there can be strong winds on the heather hills in August. In place of the queen excluder a sheet of plastic can be placed over the brood combs, leaving a one inch margin all round so that the bees gain access to the box above the plastic. This will deter the queen from ascending into the honey super.

Heather honey cannot be extracted in the same way as other honey because of its thixotropic quality. There are various ways of dealing with it. The traditional way is to cut the comb into pieces - each shallow comb cut into three slabs - and wrap it in strong scrim cloth. The honey is pressed through the cloth in a heather honey press. It is this pressing that gives ling honey its characteristic sparkle because in the pressing it collects thousands of tiny air bubbles which remain in it. A variation of this method is to use the 'Smith scraper' available from most appliance dealers. This appliance has on one side a fork with a piece of cheese wire stretched across the tines. This is pushed into the comb at one end, down to the mid-rib, or foundation, and pulled the length of the frame, undercutting the comb, first on one side and then on the other. Once all the comb has been removed the other side of the 'Smith scraper' which has a blade, is used to scrape off any remaining honey. The resultant 'potage' of honey and wax is wrapped up in scrim cloth and pressed out as before. Both of these methods are rather messy to work with.

An alternative method is to use a 'Perforextractor' which is basically a set of long, blunt needles which slide freely up and down in use. There are various hand models of this appliance and there are electrically operated ones. They work on the principle that the thixotropic ling honey can be temporarily liquefied if vigorously agitated. The combs are uncapped and the needles of the 'Perforextractor' are plunged downwards on to the comb. They enter the uncapped honey cells but do not pierce the mid-rib.

A fairly high temperature must be maintained throughout the whole process until extraction, sieving and bottling are completed. Ling honey produced in this way loses much of the sparkle which pressing imparts. The 'Perforextractor' originated in Scandinavia and is much used in these northern countries.

Many beekeepers prefer to work for cut comb honey. For this thin, unwired foundation is used. Snow white, newly built comb filled with ling heather honey is a delight to the eye. The comb can be cut in pieces of whatever size, or weight, the beekeeper may decide. Ling honey being thixotropic does not leak much when the comb is cut.

My friend, Andrew Stewart of Peebles, many years ago devised a comb cutter which is calibrated to give pieces of comb 2 oz. 4 oz. or 6 oz., as may be required. This is an excellent appliance, but unfortunately, it is not in commercial production. The cut comb can be wrapped in cling film. More attractively it can be placed in white plastic containers with clear lids. If the beekeepers cut out a template of card the exact size and shape of the container, he can then cut the comb to fit precisely, make an eye catching product.

Possibly the most effective way of producing ling heather honey is by means of 'cogs'. These are simply wooden frames made of four pieces of wood and can be of the outside dimensions of a normal super, or they can be of half or quarter size. They should be 4 1/2 ins. deep. Thin wooden slats are nailed in pairs 1 1/2 ins (38 mm) apart. Pieces of thin unwired foundation, about 2-3 ins. long and 1/2 in. deep are fixed between the slats to act as 'starters' for the bees to build comb. Some artistic beekeepers fix the starters in fanciful patterns I well remember a magnificent crown produced in the Queen's Jubilee year. The plastic sheet already mentioned may advantageously be placed under the cogs. Once the cogs are ready for removal they should be prized loose with the hive tool and a length of cheese wire affixed to a wooden handle at either end should be pulled through under the cog, severing any remaining attachment to the frames below. A small twig or matchstick is inserted at each corner and the cog left overnight for the bees to clean up the severed comb. The cog can then be placed over a clearing board, taking great care that it is bee proof. A well finished cog, of whatever size is probably the most satisfying products of beekeeping.

I have not mentioned the production of section honey. This is simply because we so seldom get favourable weather conditions for the production of sections. Bees have to be forced into sections and some beekeepers put in 'bait' section, that is at least partially built sections - but then you must first get your bait sections! Possibly a better prospect is to go for the circular 'Cobana' section which appear to be much more attractive to the bees. These circular sections require a much heavier capital outlay than the traditional square sections, but this may well be justified.

Once the heather stocks have been brought home from the hills and the honey is removed, they should be fed sugar syrup. This will encourage the bees to seal any unsealed honey, avoiding the fermentation of stores during the winter months. It also encourages the queens to continue laying for a time, providing young bees for better over-wintering. *"The best winter packing for bees is bees!"*

# Chapter Ten

It is not my intention to repeat the rather sterile arguments for and against certain types of hive. This is a problem which the hobbyist nature of beekeeping in Britain has perpetuated. In the major honey producing countries this question has been settled long ago. The hive most widely used is the Langstroth, closely followed by the Modified Dadant. These two single-walled hives are successfully used in many climes and conditions and the bees prosper and produce honey.

If colonies are to be transported, e.g., to the heather or to the fruit blossom, one must adopt one of the single-walled hives.

For brood combs one should use the Hoffman type of frames which give great stability when hives are being transported. Never go for that abomination the 'castellated' form of frame spacing which is even worse than the traditional metal spacer. Sooner or later the tin will bend or flatten. This type of spacing prevents the sliding of frames along during manipulation.

In the supers I prefer the 'Manley' close-ended frames, although some beekeepers find them more difficult to uncap. A good smoker, efficient veil and a hive tool are all essentials. If gloves gives one greater confidence, then there is no need to apologise for their use, although I have never found them necessary. I have found that a pair of wooden wedges are helpful when lifting heavy supers. As one side is levered up with the hive tool the wedges are inserted under that side of the super. One can then deal with the other sides in comfort. If the wedges are joined by a piece of string long enough to go over the beekeeper's head, when the super is carried away the wedges are also pulled away to drop by the beekeeper's sides and are ready for use again.

A useful device for carrying hives is made of two lengths of stout rope, or better still, upholsterer's webbing. Each piece should be six feet in length and have a loop at either end large enough to take a stout carrying pole. Each length of rope, or webbing should have a pole passed through both loops. (Note: the ropes are not from one pole to the other). The two poles are laid side by side and just far enough apart to permit a hive to be set down between them. The loops should overlap between the poles and the hive placed over the overlap. It may be thought that the hive will fall between them - but it doesn't and can be carried easily and safely by two people.

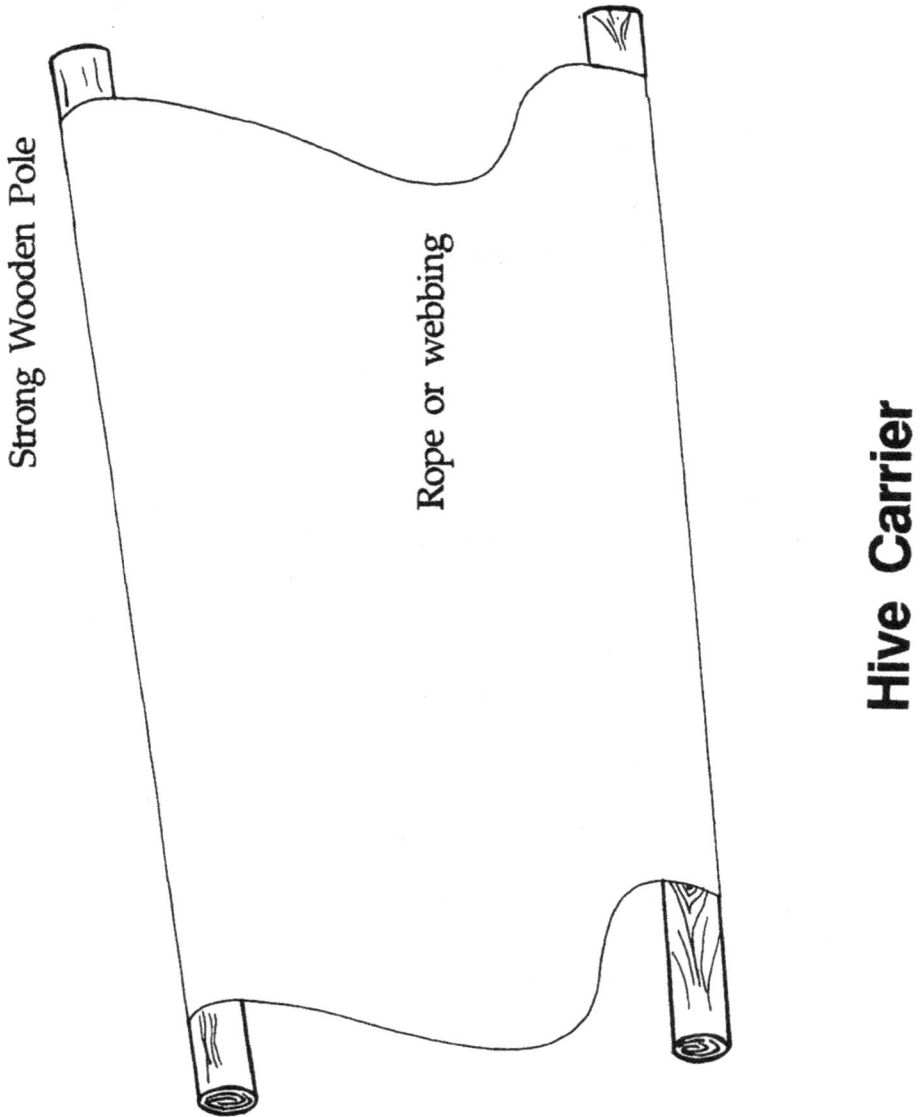

**Strong Wooden Pole**

**Rope or webbing**

## Hive Carrier

A simple hive stand can be made from two pieces of 2" x 3" wood, approx. 26" long. These should be half-checked in the middle in such a way that when fitted into each other, they form an 'X' or a St. Andrew's cross. This hive stand is particularly useful because it can easily be dismantled and re-assembled making it very convenience when taking bees to the heather. The wood should be treated with wood preservative. Wood suitable for this purpose can be obtained cheaply where old buildings are being demolished - but one must be sure that it is quite free from insect infestation.

The best type of brood frames are undoubtedly the Hoffman frames which are self-spacing and are much to be preferred to the familiar metal end - even if the metal is plastic. However, nearly forty years ago my friend Mr. J.C.V. Scherrenburg-Renkum of Holland advocated the following way to use metal ends. I quote his letter:

*"Take a metal spacer and bend the lower or upper side 135 degrees. You now have the Scherrenburg spacer. The bended lip is at the same time the spacing lip, if you wish to use the frames again in the hive with the long top bars. If this metal spacer is now put on in the right way you will find the whole on the inside of the frame, except the thickness of the metal used, thus having the whole of the lug at your disposal. You get a space, free from bees between the wall of the hive and the lip of the metal spacer."*

The late Willie Smith of Innerleithen was to comment:

*"The best way to put on the metal end is that which has been suggested by Mr. Scherrenburg, and it is so simple it is a wonder it did not dawn on someone sooner. It is to bend down one of the lips of the metal end, thereby allowing the metal end to fit close up to the side member of the frame, leaving 3/4 in. lug of frame exposed."*

# Chapter Eleven

$V$ery early in my career as a beekeeper I found my services in demand by the local authority to deal with, not only swarms, but wasp 'bikes', hornets, bumble bees, ants and even on occasion, earwigs. I must have responded, over the years, to several hundreds of such calls, but I never received a single penny for my services. Recently I listened on the T.V. to a reference to what was being charged in other areas and found that commercial firms were charging about £30 per call-out! What was most annoying was that many of the people who required my help assumed that I was being paid for doing it and treated me accordingly. In fact some had the idea that in collecting a swarm I was gaining something of great value. I remember on one occasion I was called to a woman's house (I spare her the title of 'lady') to find that she had been spraying the swarm which had clustered on her window ledge, with insecticide. Her lounge carpet was covered with dead and dying bees and only a pitiful handful of bees were clustered round the queen. I removed the small cluster and then swept up the bees from the carpet with a soft hand brush. There was nothing to do with them but to put them in the trash bin. To my astonishment, as I was about to leave, she inquired "Will you pay me now, or will you send me a cheque?" When I asked her what she meant she informed me that her son had assured her that a swarm of bees was worth a lot of money and that she was not to let me take them away without payment! I offered to return the 'valuable swarm' to her window ledge so that her son could have them, but she eventually persuaded me to take them away with me!

On another occasion I was called to deal with a swarm at a building in the outskirts of the town. In this case the bees had taken up their abode high in the chimney-head of an old three storey tenement building. Bees had been dropping down the chimney, but the fire-place had been blocked up. Naturally they made for the light which showed below the skirting-board and were crawling over the carpets of rooms occupied by a young couple and their baby. The fire brigade had been summoned at the same time as I was and we were discussing how to deal with the situation when the minister of a nearby church arrived on his bicycle. *"I hear you have got my swarm"* he exclaimed. I asked how he knew that it was his swarm and he informed me that his housekeeper had watched it fly in this direction the day before. I reminded him that the law said that whoever found a swarm was entitled to keep it. *"Oh, but I know you would not do that to me, I cannot afford to lose a swarm."* I assured him that I had no intention of depriving him of his swarm and I then pointed out the situation to me. *"You know,"* he said *"I do not think these are my bees at all - I'll just leave them."* With that he rode off.

It was too big a job to open up the chimney head so we smoked as many of the bees out as possible and sealed the skirting boards thoroughly. I have never kept any of the swarms I retrieved. I put them in an out apiary until I was sure that they were disease free, and reasonable to manage and there was always a beginner who wanted a start in beekeeping.

This work had its humorous side too. I was once called to the home of an elderly maiden lady to find that she had shut herself in her kitchen and that all the other doors of rooms were draped with sheets, curtains or blankets *"to keep the bees away."* She told me that she had arrived back home from holiday the evening before and had been so tired that she had gone to bed at once. She awoke early in the morning to find bees swirling all round her bed and her bedroom window *"black with bees."*Sure enough, when I looked into the bedroom there were a lot of bees. They were entering and leaving by way of a ventilator fan which had been fitted to one of the window panes. Then I noticed that the bees were flying under her bed. I looked under the bed and there was a chamber pot which had been put upside down on a sheet of newspaper. I turned it over and there it was, filled to the brim with newly built comb and complete with several slabs of brood. At my request the lady furnished me with some lengths of sewing tape and a spatula knife from the kitchen. I had some empty brood frames and a nucleus box in my car. Equipped with these I carefully cut all the comb out of the bedroom receptacle; fitted the pieces into the brood frames and tied them up with the tape. Soon they were all safely in the nuc box. I removed the chamber pot and told the lady that I would leave the nuc box in her bedroom until dusk. In the evening I returned, closed the nuc box and put it in my car. I was invited in for a cup of coffee and as we sat chatting she suddenly began to laugh. I asked what was the cause of her merriment. *"I have been thinking that it is just as well that I did not use that pot last night!"* she said. *"Just as well for you - and for the bees!"* was my reply .

One of the swarms that called for some ingenuity in its capture. had settled on the roof of a large greenhouse, the structure of which was very unsound. There was no way one could reach the cluster. After a good bit of cogitation, my helper and I rigged up a rope and pulley arrangement. We firmly fixed two well-used brood combs inside a cardboard carton and managed to manoeuvre it right over the cluster and gently lowered it down on to the glass roof. We left it in position until dusk and then very gingerly reeled it in. The bees had, obligingly, taken possession of the two combs and the carton. After that it was easy to install them in a hive..

Swarming bees can provoke some amusing situations. A bee-keeper friend of mine was session clerk of a country church and when the minister went off on his annual holiday, somehow the session clerk managed to obtain beekeeping parsons to act in the interval. On one such occasion the visiting minister was invited to lunch by my friend. They were in the middle of the meal when one of his hives swarmed. Soon both minister and session clerk were up a tree dealing with the swarm - Sunday notwithstanding. Once they had captured it they proceed to try to find out from which hive it had emerged. To their dismay they found queen cells in each hive they opened. Time sped on and soon they were startled to hear the hostess's voice call. *"Look you two, your evening service is due in fifteen minutes!"* A hurried wash and brush up and they dashed down to the church. Alas when the minister got into the vestry he found that there were several bees crawling up inside his trouser legs. He stamped his feet but failed to dislodge the unwelcome explorers. Nothing for it. He quickly removed his trousers, turned them outside in and commenced to pick the offending bees off. In the midst of doing this he was horrified to hear the sound of two pairs of feminine feet approaching the vestry door and a female voice saying *"We will give it to him before he goes into church."* A hurried look round the unfamiliar room revealed only one other door. He quickly grabbed his trousers and stepped through the door - to find himself in full view of the congregation!

Beekeepers are no different to other members of society. you get all types. One lady I knew had embroidered top covers for each hives with illustrations of the main plants important to the honey bee. One other lady in her eagerness to get a good crop of honey had extracted every comb in the hive. Next day she phoned to say that her honey had *"gone all maggoty."*

I have rather less pleasant memories of a young farmer and his wife. I had given the couple a complete hive and excellent colony as a wedding gift - he had told me that he wanted to be a beekeeper. A few weeks later he telephoned to say that he had fallen heir to five colonies belonging to his wife's uncle. *"They have been neglected for years"* he said *"and the hives are in a bad state. I have about five new National hives, will you come and help me to transfer the bees?"* Of course I agreed and set off about four o'clock in the afternoon. He was not much good at helping and it was back-breaking work on a hot afternoon. Frames were jammed together with propolis and combs had broken down and been fused together by the bees. Each frame had to be levered out with the hive tool and many had to be discarded. In the midst of my labours his wife called him away and I concluded that some duty concerning the cattle had to be attended to. However when he returned he was chewing and there was a distinct odour of bacon and eggs!

It was after eight o'clock when I finally finished. I had to ask to get washing my hands before I drove home. I was shown into an outhouse and the only towel offered was a piece of sacking. There were two large trays piled high with eggs. *"Would you like some eggs to take home to your wife?"* He inquired. "That would be very nice" I said. Two dozen eggs were put into a paper bag and put into my car. I was about to drive off when he said *"That will be (so much) for the eggs"*. I silently paid him and drove off. He was on the phone a week later *"Will you come out and show me what to do next?"* I suddenly found that I had a very important engagement!

Nevertheless I have to thank my beekeeping for the many good friends I have made over the years and in so many countries.

www.ingramcontent.com/pod-product-compliance
Lightning Source LLC
Chambersburg PA
CBHW081423270326
41931CB00015B/3382

# X-CELSIOR

**THE UNOFFICIAL AND UNAUTHORISED
GUIDE TO THE MARVEL MULTIVERSE
UPDATED FOR 2023**

# CONTENTS

## INFO BURSTS

# INTRODUCTION

Once, there was a book called *The Unofficial And Unauthorised Guide To The Movies Of Marvel*. When the Marvel Cinematic Universe continued unabated, it was split into two sets of guides – *It's All Connected*, the unofficial guide to the MCU, and *X-Celsior*, the unofficial guide to all the other stuff that wasn't part of the MCU. Everything was going swimmingly, until Disney bought Fox, and then Andrew Garfield and Tobey Maguire turned up in *SPIDER-MAN: NO WAY HOME*. Suddenly, it looked like *X-Celsior* was now the guide to the multiverse, beyond the MCU. It's been quite the ride writing these books, believe me.

So, in your hands, you hold the latest update to *X-Celsior*. In this book, you'll find the movies that are made by Fox Studios, Sony, Colombia and every other studio that's briefly had their hands on a Marvel property (or still do, in the case of Sony). While the movies have a great deal of depth to them, you'll also find a quick, but hopefully useful look, at the television series outside the MCU that have been made over the years using Marvel characters. There are summaries, cast listings, comic book comparisons and more for every movie, with a few info bursts for additional information. In this book you'll find the X-Men and the Fantastic Four, along with the earlier adventures of Peter Parker, Matt Murdock and Howard the Duck. You will not, however, find the Tony Stark, Jessica Jones or Cloak & Dagger. This is the place to find *LOGAN*, but not *THOR*. However, you will find *Legion* here, even though it's co-produced by Marvel Television. More on that later when you get to it.

*X-Celsior* contains a lot of information, but with Marvel regaining all their properties, there's less and less that is coming from outside Marvel studios. As such, there's not a huge amount of new information coming each year. So, whilst this book is updated and revised, with a lot of new information, in future there'll be very little to update. As such, the next edition of this book may not be for a few years, and will indeed be another updated/revised edition. Additionally, if you have *It's All Connected* you might notice this introduction is broadly similar to the introduction there, and the history of Marvel is almost identical. Don't worry – that's the only things that are common between the two. You haven't wasted money if you bought both.

If you bought *The Unofficial And Unauthorised Guide To The Movies Of Marvel* all those years ago, these books have evolved so much, that this new purchase will definitely be worthwhile (I'd also like to thank you from the bottom of my heart for your support). If this is the only Marvel guide you've picked up of mine, I hope you'll also feel you haven't wasted your money. I enjoy very much writing these books (not least because I love Marvel's output and this gives me an excuse to watch them more than once), so if your enjoyment in reading equals mine in writing, then this will have been an excellent exchange. Finally, I'd like to thank the following people for the help in writing this book: Kevin, who I've discussed Marvel with endlessly; Andrew, who gave me a lot of assistance on the original cover of the first book; Comics Etc, the best comic shop in Brisbane – buy from them always; Clinton and Bryan who bought copies of the book as soon as it was published, and wanted me to sign them... why, I can't even imagine; Deb, who bought several copies of the original book, and was determined to make it succeed; Michael and Aaron, my two great friends who I've spent far too long discussing all sorts of things with, and who have been unswerving in their support of this book; Jessica, who is one of the greatest people on the planet and constantly promotes my work, which I can't thank her enough for; my sister Brooke, who bought a copy and dutifully read it; my Mum, who bought a copy and didn't read it; my Dad who didn't buy a copy and obviously didn't read it, but is a champ in his own way; Fynn, Niko and Remy, who think because I've self-published a book, I'm some sort of famous and amazing author...kids, eh?

And to everyone who has bought this book, or its predecessors, thanks. I won't make a fortune with these books, nor will I become famous, but the fact it sold to the people it did has made it all worthwhile, so thanks a lot for that.

# HOW TO USE THE BOOK...

This book principally looks at the movies that exist outside of the Marvel Cinematic Universe; ie the films that are not made by Marvel Studios. As such, section one is a list of live action films, section two is a list of live action television films, while section three are animated films.

The fourth and fifth sections exists more for the sake of completeness than anything else, and are an overview of Marvel based television programs, animated and live action. These entries are less detailed than the movie entries, though hopefully still full of useful information.

Each entry uses the following format:

### Title

***(Date of initial release – Original television station where applicable)***
Short summary of the story.

**Cast:** For television shows this is the regular credited cast. For movies it's generally the cast that feature in the credits sequence. Cast and characters that are italicised are characters lifted from the comics.

**Crew:** The production crew involved. **Writers, Directors, Producers, Composers, Executive Producers, Directors of Photography, Production Designers, Editors, Production Companies** all where applicable.

**Notes:** Background notes on the movie, and general points of interest. This book tends to avoid unsubstantiated rumour, but where there are recorded details of back stage controversy, it's mentioned.

**Stan Spotting:** Where to look out for the cameo appearance of Stan "The Man" Lee. If there's no entry, it's because the company took the frankly ludicrous decision not to include him.

**Comic Notes:** A look at the comic versions of movie characters. As the book goes on these get smaller, as characters will be covered in their first appearance. For comic fans, this will provide background details; for those less that way inclined, hopefully it will provide an interesting look at the different path the movies take.

**Ratings:** The ratings from IMDb, Rotten Tomatoes and Metacritic, where available.

**Review:** The author's own personal opinion on the media. If it's not present, it simply means I haven't seen it, and don't want to give an uniformed opinion.

# CHRONOLOGICAL LIST OF MARVEL FILMS

**1944**
Captain America *(Republic Pictures)*

**1986**
Howard the Duck *(LucasFilm)*

**1989**
The Punisher *(Live Entertainment)*

**1990**
Captain America *(21st Century Film Corporation)*

**1998**
Blade *(New Line)*

**2000**
X-Men *(20th Century Fox)*

**2002**
Blade II *(New Line)*
Spider-Man *(Colombia Pictures/Sony)*

**2003**
Daredevil *(20th Century Fox)*
X2 *(20th Century Fox)*
Hulk *(Universal Studios)*

**2004**
The Punisher *(Artisan Entertainment)*
Spider-Man 2 *(Colombia Pictures/ Sony)*
Blade Trinity *(New Line)*

**2005**
Elektra *(20th Century Fox)*
Man-Thing *(Lionsgate)*
Fantastic Four *(20th Century Fox)*

**2006**
X-Men: The Last Stand *(20th Century Fox)*

**2007**
Ghost Rider *(Colombia Pictures)*

Fantastic Four: Rise of the Silver Surfer *(20th Century Fox)*

**2008**
Iron Man *(Marvel Studios)*
The Incredible Hulk *(Marvel/Universal)*
Punisher: War Zone *(Marvel Studios)*

**2009**
X-Men Origins: Wolverine *(20th Century Fox)*

**2010**
Iron Man 2 *(Marvel Studios)*

**2011**
Thor *(Marvel Studios)*
X-Men: First Class *(20th Century Fox)*
Captain America *(Marvel Studios)*

**2012**
Ghost Rider: Spirit of Vengeance *(Colombia Pictures)*
The Avengers *(Marvel Studios)*
The Amazing Spider-Man *(Colombia Pictures)*

**2013**
Iron Man Three *(Marvel Studios)*
The Wolverine *(20th Century Fox)*
Thor: The Dark World *(Marvel Studios)*

**2014**
Captain America: The Winter Soldier *(Marvel Studios)*
The Amazing Spider-Man 2 *(Colombia Pictures)*
X-Men: Days of Future Past *(20th Century Fox)*
Guardians of the Galaxy *(Marvel Studios)*

## 2015

Avengers: Age of Ultron *(Marvel Studios)*
Ant-Man *(Marvel Studios)*
Fantastic Four *(20th Century Fox)*

## 2016

Deadpool *(20th Century Fox)*
Captain America: Civil War *(Marvel Studios)*
X-Men: Apocalypse *(20th Century Fox)*
Doctor Strange *(Marvel Studios)*

## 2017

Logan *(20th Century Fox)*
Guardians of the Galaxy Vol 2 *(Marvel Studios)*
Spider-Man: Homecoming *(Marvel Studios/Columbia Studios)*
Thor: Ragnarok *(Marvel Studios)*

## 2018

Black Panther *(Marvel Studios)*
Avengers: Infinity War *(Marvel Studios)*
Deadpool 2 *(20th Century Fox)*
Ant-Man and the Wasp *(Marvel Studios)*
Venom *(Sony)*

## 2019

Captain Marvel *(Marvel Studios)*
Avengers: Endgame *(Marvel Studios)*
Dark Phoenix *(20th Century Fox)*
Spider-Man: Far From Home *(Marvel Studios/Columbia Studios)*

## 2020

The New Mutants *(20th Century Studios)*

## 2021

Black Widow *(Marvel Studios)*
Shang-Chi And The Legend Of The Ten Rings *(Marvel Studios)*
Venom: Let There Be Carnage *(Columbia Studios)*
Eternals *(Marvel Studios)*
Spider-Man: No Way Home *(Marvel Studios/Columbia Studios)*

## 2022

Morbius *(Columbia Studios)*
Doctor Strange In The Multiverse Of Madness *(Marvel Studios)*
Thor: Love And Thunder *(Marvel Studios)*
Black Panther: Wakanda Forever *(Marvel Studios)*

## 2023

Ant-Man And The Wasp: Quantumania *(Marvel Studios)*
Guardians Of The Galaxy Vol 3 *(Marvel Studios)*
The Marvels *(Marvel Studios)*

# A BRIEF HISTORY OF MARVEL

In 1939, Martin Goodman founded a comics company called Timely Publications. Its first comic was *Marvel Comics #1* which featured the (original) Human Torch and Namor, the Sub-Mariner. Two years later *Captain America Comics #1* debuted, and it's easy to guess who head-lined that particular comic. Both these comics sold around 900,000 copies, proving to be instant successes. From the outset Goodman had hired his wife's cousin – Stanley Leiber – to do general work around the office, though this later saw him do some writing under a name that would become better known over time: Stan Lee.

After the war, super heroes lost favour with the general public, and in November, 1951, Timely Publications became Atlas Comics, and although Goodman tried to resurrect the Human Torch, Namor and Captain America, none of them were particularly well received. However, Atlas managed to survive comfortably without them.

In June, 1961 Atlas changed its name to Marvel Comics, and as Detective Comics (or DC) had success in resurrecting superheroes, Marvel decided to follow suit. Under the editorial eye of Stan Lee, November 1961 saw the debut of The Fantastic Four, and from there the company continued to expand.

Perhaps what most notably separated Marvel Comics from DC, was, initially, a greater tie to the "real" world. Unlike DC, which created cities such as Gotham and Metropolis, Marvel's superheroes lived – for the most part – in real cities, predominantly New York. In addition to this, unlike DC's heroes who were, effectively, gods walking among us, Marvel's heroes were – to paraphrase – ordinary people doing extraordinary things. They were the everyman (for the most part), given unexpected powers and abilities, and struggling to cope with these; struggling to do the right thing, even though their own human failings made that sometimes difficult.

Marvel's titles did solid business throughout the seventies and eighties, though they were unable to topple DC from the top shelf. However, in the early nineties, comic books became big business and under Editor in Chief Tom DeFalco, Marvel capitalised on this and crushed its opposition, to the point where Warner Bros who now owned DC Comics approached Marvel with the offer to license DC's characters – an offer that was refused.

But all good things come to an end, and as the mid-nineties hit and the boom busted, Marvel fell harder than most, declaring bankruptcy in 1996. Two years later, Toy Biz and Marvel Entertainment Group solved the bankruptcy issue by forming Marvel Enterprises (which, in 2005, changed its name to Marvel Entertainment). Isaac Perlmutter – the co-owner of Toy Biz –

and his partner Avi Arad, with Marvel's publisher Bill Jemas and new editor in chief Bob Harras started to turn the tide of Marvel's fortunes. One of the first things they did was to start selling off the film rights to Marvel's various properties.

Marvel had actually licensed its properties as far back as the forties, when Captain America was licensed to Republic Pictures. From that point on very little happened with the characters in a visual medium, until the late seventies when the Marvel Entertainment Group was formed, and licensed out a number of properties for development, though mostly as television specials. While DC and their sister company Warner Bros had been releasing Superman movies during the late seventies and eighties (admittedly to diminishing returns), the best Marvel could do was **Howard the Duck**, a 1986 flop. In 1989 Warner Bros released the critically acclaimed **Batman**, while Marvel Entertainment produced another flop with **The Punisher**. With the need to escape bankruptcy, Marvel started to look towards licencing their products as movies to gain money.

In 1993, Marvel Films was formed, headed by Avi Arad. Marvel Entertainment co-produced many of the movies that were distributed by other companies, but Marvel Films itself made nothing at all. In 1996 the name was changed to Marvel Studios and Arad – as head of Marvel Films – and Jerry Calabrese, the president of Marvel Entertainment – effectively became joint CEO's of Marvel Studios. Marvel Studios held responsibility for licensing productions and Marvel's products were seized upon, almost vulture like. New Line Cinema, Columbia, Twentieth Century Fox, Lions Gate and Paramount Pictures all negotiated deals, but while the deals were generally not that great for Marvel, the company was cunning enough to include a "use it or lose it" clause, ensuring that if a movie wasn't made within a certain time, the rights for those properties would return to Marvel. In 2004, Marvel Studio's COO, David Maisel, was advised that Marvel should actually produce their own films, rather than licence the production. Maisel managed to obtain funding ($525 million, in fact, for ten films over eight years), by giving up ten of its characters rights as collateral. Having got the rights back for Iron Man and the Hulk, Marvel released its first two films in 2008, under the control of the new President of Production, Kevin Feige.

New Line Cinema had some success with **Blade**, and Twentieth Century Fox had even more success with **X-Men**. Columbia's **Spider-Man** was a financial and critical hit, with only Lion's Gate's **The Punisher** not getting the success that everyone expected. However, all of that was eclipsed by the success of **Iron Man**, which, while it may not have made as much money as others, was expected to fail. Feige's long term plans for his movies, and a decision to effectively treat them as a television series with him as the showrunner, saw a surprising new trope emerge – that of a connected universe. As companies started to lose their rights back to Marvel, Marvel Studios continued to build and develop its Cinematic Universe in phases. The MCU has gone from being a surprise success, to one of the most powerful film studios in Hollywood. So much so, that Disney bought Marvel out in 2009, and latterly Fox Studios in 2018. Universal still holds the distribution rights for the Hulk (which is why there won't be any solo Hulk films until that contract expires), and the rights for Namor the Sub-Mariner remain uncertain (but seem to be similar to Hulk). Sony, of course, have Spider-Man and all that entails, and currently remain in a strange relationship with Disney/Marvel.

## Marvel Comics' Publishers

Abraham Goodman (1939 – 1940)
Martin Goodman (1940 - 1972)
Charles "Chip" Goodman (1972)
Stan Lee (1972 – 1996)
Shirrel Rhoades (1996 – 1998)
Winston Fowlkes (1998 – 1999)
Bill Jemas (2000 – 2003)
Dan Buckley (2003 – 2017)
John Nee (2018 – Incumbent)

## Marvel Comics' Editors-in-Chief

Martin Goodman (1939 – 1940)
Joe Simon (1940 – 1941)
Stan Lee (1941 – 1942)
Vincent Fargo (1942 – 1945) *NB Acted for Stan Lee, who was in the war.*
Stan Lee (1945 – 1972)
Roy Thomas (1972 – 1974)
Len Wein (1974 – 1975)
Marv Wolfman (1975 – 1976)
Gerry Conway (1976)
Archie Goodwin (1976 – 1978)
Jim Shooter (1978 – 1987)
Tom DeFalco (1987 – 1994)
Bob Harras (1995 – 2000)
Joe Quesada (2000 – 2011)

Axel Alonso (2011 – 2017)
CB Cebulski (2017 – incumbent)

## Marvel Studios Officers

President: Kevin Feige
Co-President: Louis D'Esposito
Parliament: Stephen Broussard, Eric Hauserman Carroll, Nate Moore, Jonathan Schwartz, Trinh Tran, Brad Winderbaum
Production and Development: Sana Amanat, Grant Curtis, Chris Gary, Brian Gay, Kevin R Wright

# I. LIVE ACTION FILMS

# CAPTAIN AMERICA
## *(5/2/1944)*

The Scarab – actually archaeologist Cyrus Maldor – develops the Dynamic Vibrator (yes, really) as a weapon, along with using something called the Purple Death, a chemical which causes a hypnotic state that makes people commit suicide. The Mayor orders Commissioner Dryden and DA Gardner to solve the crime, and Gardner goes undercover as Captain America, to confront Maldor.

**Cast:** *Dick Purcell (Captain America/Grant Gardner)*, Lorna Gray (Gail Richards), Lionel Atwill (Dr Maldor), Charles Trowbridge (Commissioner Dryden), Russell Hicks (Mayor Randolph), George J Lewis (Matson), John Davidson (Gruber), Stanley Price (Chemist)

**Dir:** Elmer Clifton, John English; **Writers:** Royal Cole, Harry Fraser, Joseph Poland, Ronald Davidson, Basil Dickey, Jesse Duffy, Grant Nelson; **Prod:** William J O'Sullivan; **Music:** Mort Glickman; **DOP:** John MacBurnie; **Ed:** Wallace Grissell, Earl Turner; **Costumes:** Unknown; **Republic Pictures**; 243 (15 chapters); $182K; Box Office unknown

**Notes:** In actual fact this was a fifteen part serial shown at theatres. Aside from the name Captain America, there is absolutely no connection to the comics whatsoever. Indeed, Captain America doesn't even have a shield. The serial was, however, the most expensive of its kind ever produced. Curiously the comics often paid a homage to the serial, even going so far as to suggest the serial exists within Marvel continuity. The episode titles were: The Purple Death; Mechanical Executioner; The Scarlet Shroud; Preview of Murder; Blade of Wrath; Vault of Vengeance; Wholesale Destruction; Cremation in the Clouds; Triple Tragedy; The Avenging Corpse; The Dead Man Returns; Horror On The Highway; Skyscraper Plunge; The Scarab Strikes; The Toll Of Doom.

**Ratings:** *IMDB:* 61%

**Review:** From a comics perspective this is a huge disappointment, with the movie sharing virtually nothing with its comic counterpart. Additionally, you have to view this within the context of its production – this is a very 1940's serialised drama. If you don't like that sort of thing, you really won't like this. But that aside – and some very dodgy acting and special effects – the movie isn't all that bad.

# HOWARD THE DUCK
## *(1/8/1986)*

Howard T Duck is sucked out of his universe and finds himself befriending Beverly, lead singer of a group called Cherry Bomb. With Beverly's help, and a friend of hers who works with a scientist, Howard tries to return to his home. As the scientist created a dimension jump device, the group attempt to reverse the process, but only end up summoning the Dark Overlord to Earth.

**Cast:** *Lea Thompson (Beverly Switzler)*, Jeffrey Jones (Dr Walter Jenning), Tim Robbins (Phil Blumburtt), *Ed Gale (Howard T Duck)*, Chip Zien (Howard's voice), *Tim Rose (Howard T Duck), Steve Sleap (Howard T Duck), Peter Baird (Howard T Duck), Mary Wells (Howard T Duck), Lisa Sturz (Howard T Duck)*, Paul Guilfoyle (Lt Welker), Tommy Swerdlow (Ginger Moss)

**Dir:** Willard Huyck; **Writers:** Willard Huyck, Gloria Katz; **Prod:** Gloria Katz; **Music:** John Barry; **Exec.Prod:** George Lucas; **DOP:** Richard H Kline ASC; **Prod.Des.:** Peter Jamison; **Ed:** Michael Chandler, Sidney Wolinsky; **Costumes:** Joe Tompkins; **Universal Pictures/Lucasfilm**; 111; c $33m; $38m

**Notes:** As indicated by the credits, a large number of actors donned the Howard costume to play the duck. Sylvester Levay provided additional music for the film when it was decided that John Barry's score sounded too old fashioned, and Thomas Dolby wrote the songs used, and picked the members of Cherry Bomb. Thompson and the other Cherry Bomb members (Dominique Davalos as Cal, Liz Sagal as Ronette, and Holly Robinson as KC) were all musicians, and performed the songs Cherry Bomb performs. Thompson, in fact, claims she still has the guitar she used in the movie.

**Comic Notes:** *Howard*: first appeared in *Adventure into Fear #19 (Dec, '73)* and was created by Steve Gerber and Val Mayerik (though only Gerber is credited as Howard's creator in the movie). There is a strong metafictional element to his comics, which tends to ensure Howard doesn't mix a great deal with the mainstream Marvel characters. His background is not quite what is depicted in the movie; in the comics it was Thog the Nether-Spawn that took him from his universe, and had him encounter Beverly Switzler. Over time he has actually encountered a large number of characters, and had a tumultuous relationship with Beverly that has, nonetheless, lasted for the most part. He did fight the Skrulls in *Secret Invasion*, and he and Beverly attempted to register as superheroes in *Civil War*. He continues to remains a part of the mainstream universe through various cameos.

*Beverly Switzler:* first appeared in *Howard the Duck #1 (Jan, '76)* and was created by Gerber and Frank Brunner. Unlike her movie counterpart, Beverly was not a singer, rather she was a nude model. Most of her storylines are tied in closely to Howard's, thanks to her on-again, off-again relationship with Howard, though she was one of the many applicants vying to babysit Luke Cage and Jessica Jones' daughter, Danielle. Curiously she also gained the ability to fly, thanks to being coated in a pink substance, but it's not a power she uses often.

*And:* the entire film was adapted for *Marvel Super Special #41 (Nov, '86)*, written by Danny Fingeroth, with art by Kyle Baker. The following month, *Howard the Duck: The Movie* three part limited series debuted.

**Ratings:** *IMDB:* 45%; *Rotten Tomatoes:* 14%; *Metacritic:* 28

**Review:** Quite why anyone thought that the first Marvel film in 40 years should use one of the most obscure characters ever created is anyone's guess. The film is so completely out there it's hard to dislike it, but when duck boobs and interspecies relationships raise their heads, prepare to ask some questions. Oddly, as soon as Tim Robbins turns up the film loses its way, before Jeffrey Jones arrives to add to the madness, but strangely bring the movie back on track. Lea Thompson deserves a medal for her outstanding performance.

## THE PUNISHER
### *(5/10/1989)*

The Yakuza attempt to make a move on the Mafia families by kidnapping the children of the important family members. Gianni Franco finds himself joining forces with the Punisher – a former cop whose family Franco had killed, and who is now taking out the criminal underworld.

**Cast:** *Dolph Lundgren (Frank Castle/The Punisher),* Louis Gossett Jr (Jake Berkowitz), Jeroen Krabbe (Gianni Franco), Kim Miyori (Lady Tanaka), Bryan Marshall (Dino Moretti), Nancy Everhard (Sam Leary) and Barry Otto as Shake

**Dir:** Mark Goldblatt; **Writer:** Boaz Yakin; **Prod:** Robert Mark Kamen; **Music:** Dennis Dreith; **Exec.Prod:** Robert Guralnick; **DOP:** Ian Baker ASC; **Prod.Des.:** Norma Moriceau; **Ed:** Tim Wellburn; **Costumes:** Norma Moriceau; **Live Entertainment**; 89; $9m; $30m

**Notes:** Stan Lee is billed as an "executive consultant" on the film. The film was made in Australia, and given a general worldwide cinematic release, though not, curiously, in America.

**Comic Notes:** *Frank Castle:* Created by Gerry Conway, Ross Andru and John Romita Sr, Francis Castiglione, as he was actually named at birth, first appeared in *The Amazing Spider-Man #129 (Feb, '74).* Castle traditionally had his family killed by mobsters, and this created his drive to initially get revenge on those that killed his family, and then to continue to pursue justice where the criminal justice system couldn't or wouldn't. Obviously his first run in was with Spider-Man, but Frank Miller frequently used him against (and with) Daredevil, and over time the Punisher has teamed up with a number of heroes, as well as briefly joining Heroes For Hire and the Secret Defenders.

    *And:* this movie takes great pains not to use any elements of the comics outside the central character, which makes this a fairly short entry all told.

**Ratings:** *IMDB:* 56%; *Rotten Tomatoes:* 28%

**Review:** It's hard to be positive about this film as it's quite a mess. Lip service is paid to the comic book character, as this Punisher is far more influenced by the movies that were being released at the time – **The Punisher** is less a comic book movie, and more a standard eighties cop flick, and surprisingly it does tend to work on that level; though Lundgren is so charmless, the likes of Gossett Jr have to work extra hard when he's in the scene. Worth watching, though not a credit to Marvel.

## CAPTAIN AMERICA
### *(14/12/1990)*

In 1936 a young prodigy is taken and experimented on, turning him into the Red Skull. Seven years later the same doctor, having defected, performs a similar experiment on Steve Rogers, turning him into Captain America. When Rogers and the Skull battle, Rogers loses and in an effort to stop a missile hitting the White House, to which he is attached, Rogers finds himself trapped in Alaskan ice for forty-seven years. When he awakens, he finds the Skull is still alive; an American gang lord, who is about to kidnap the President in order to stop him from implementing an ecological program that will cost manufacturers millions while saving the planet.

**Cast:** *Matt Salinger (Steve Rogers/Captain America),* Ronny Cox (President Tom Kimball), Ned Beatty (Sam Kolawetz), Darren McGavin (General

Fleming), Michael Nouri (Lt Col Louis), Melinda Dillon (Mrs Rogers), *Francesca Neri (Valentina de Santis),* Bill Mumy (Young Fleming) introducing *Kim Gillingham (Bernie Stewart/Sharon) and Scott Paulin as "Red Skull"*

**Dir:** Albert Pyun; **Writer:** Stephen Tolkin, Lawrence J Block; **Prod:** Menaham Golan; **Music:** Barry Goldberg; **Exec.Prod:** Stan Lee, Joseph Calamari; **DOP:** Philip Alan Waters; **Prod.Des.:** Douglas Leonard; **Ed:** Jon Poll; **Costumes:** Heidi Kaczenski; **Marvel Entertainment/Jadran Film**; 97; $3m; Unknown

**Notes:** Originally intended for release to coincide with the fiftieth anniversary of Captain America, the film did get some international release, but in the US was held back for a straight-to-video release in 1992. The film provides a vastly different background for the Red Skull (now an Italian), and a number of names are changed in the film. Interestingly, both the Sub-Mariner and the (original) Human Torch are mentioned in the film.

**Comic Notes:** *Steve Rogers*: Joe Simon and Jack Kirby created Captain America for *Captain America Comics #1 (Mar, '41),* making him the oldest Avenger in every sense of the word. His comics' origin is broadly similar to what is portrayed in this film, though he never flew over the White House in any comic version. When he recovered he almost immediately joined the Avengers, which, of course, is not the case in the film. The mantle of Captain America has been taken on by a number of different people, both after Rogers' burial in ice, and since his return, when he has been forced to give the job up.

*The Red Skull*: or Johann Schmidt; debuted in the same series six issues later *(#7 Oct, '41),* created by the same team, with France Herron. His backstory in the comics is very different to the movie – he was German, and though he was an early test subject for the super soldier serum, he continued as one of Hitler's generals throughout the war. The Skull returned, similar to Steve Rogers, to visit cruelty and violence against various superheroes, concentrating on Captain American and the Avengers, but the movie's plot is something he has never been involved in.

*Sin:* The Red Skull's daughter first appeared in *Captain America #290 (Feb, '84).* She was created by J M DeMatteis and Paul Neary and is named Sinthea Schmidt, or Sin for short. Due to the change in name for the Red Skull, in this movie she becomes Valentina de Santis.

*Bernie Rosenthal*: (not Stewart) made her debut in *Captain America #247 (Jul, '80)* and was created by Roger Stern and John Byrne. In the comics she is a lawyer who ultimately becomes engaged to Steve Rogers.

*And:* Sharon is, of course, named for Sharon Carter.

**Ratings:** *IMDB:* 33%; *Rotten Tomatoes:* 9%

**Review: Captain America** does itself few favours. From an opening where the majority of the Italian dialogue is not subtitled, the movie then moves forward to establish Captain America in a World War 2 setting, before skipping forward to have the 90's sequences. The problem with this is, too much is skipped over and too much is unbelievable. We never really believe Captain America becomes a symbol, or indeed that he is an effective soldier. It's not helped when the best performances come from Cox, Beatty and Gillingham in their surprisingly far more heroic roles. Even Neri and Paulin seem more interested in the film than Salinger. The highlight of the film is the idea that the Red Skull's daughter is a Paris Hilton type, complete with entourage. The script is far too flabby, trying to tell too many stories, and not telling any effectively.

## THE FANTASTIC FOUR
### (N/A)

Reed Richards and Victor Von Doom attempt to harness the power of the Colossus comet, but the result scars Victor so badly, Reed believes him dead. Ten years later, Reed obtains an enormous diamond to recreate the experiment, but the Jeweller steals the diamond and Reed and his crew obtain super powers. But Doom is waiting to get the diamond and have his revenge.

**Cast:** *Alex Hyde-White (Dr Reed Richards), Jay Underwood (Johnny Storm), Rebecca Staab (Susan Storm), Michael Bailey Smith (Ben Grimm),* Ian Trigger (Jeweller), *Joseph Culp (Dr Victor Von Doom),* George Gaynes (Professor), *Kat Green (Alicia Masters), Carl Ciarfalio (Thing),* Charles Butto (Weasel), Annie Gagen (Mrs Storm), Howard Shangraw (Kragstadt), David Keith Miller (Trigorin), Robert Beuth (Dr Hauptman), *Mercedes McNab (Young Susan), Philip Van Dyke (Young Johnny)*

**Dir:** Oley Sassone; **Writer:** Craig J Nevius, Kevin Rock; **Prod:** Steven Rabiner; **Music:** David Wurst, Eric Wurst; **Exec.Prod:** Roger Corman, Bernd Eichinger; **DOP:** Mark Parry; **Prod.Des.:** Mick Strawn; **Ed:** Glenn Garland; **Costumes:** Reve Richards; **Constantin Film Produktion/New Horizons**; 90; $1m; N/A

**Notes:** Reputedly this movie was never intended for release and was simply an ashcan film in order for Constantin Film Produktion to retain the rights, which they did. However, Corman and Eichinger argue this was not the case, and that they had every intention of releasing the film. The film's budget was allegedly $1,000,000 and Marvel film chief (at the time) Avi Arad bought the film for twice that amount in order it was not released, for fear it would damage the brand. It is possible to track the film down, however.

**Comic Notes:** *The Fantastic Four:* All four members first appear – surprisingly – in *The Fantastic Four #1 (Nov, '61)*, and were created by Stan Lee and Jack Kirby. Their powers are all replicated correctly in this movie (also surprisingly), though over time in the comics, Sue has been able to use her powers as a force field to allow her to protect herself, or even fly. The four have had a huge variety of storylines over the years, seeing Reed and Sue get married, have two children (Franklin and Valeria), and get involved in a love triangle with Namor, the Sub-Mariner, and also, rather bizarrely, Dr Doom (indeed, Valeria's parentage is curiously questionable thanks to Valeria also being the child of Sue and Doom in the future). *Civil War* split the team in half, and then split Reed and Sue as they took different sides, and Sue and Johnny have both died and been brought back. Their headquarters has changed from the Baxter Building to Four Freedoms Plaza, and the line-up has occasionally changed, including such heroes as Spider-Man, Luke Cage, Crystal, Storm, Black Panther, Ms Thing, Nova and She-Hulk. At one point the team even changed its name to the Future Foundation (getting white costumes in the process). At present the Fantastic Four has disbanded, with the Thing a member of the Guardians of the Galaxy, Human Torch working with the Inhumans (alongside his former girlfriend Crystal), and the Richards family attempting to rebuild the multiverse.

*Doctor Victor Von Doom:* first appeared in *The Fantastic Four #5 (Jul, '62)*, and is again much the same as his movie appearance, though his armour does give him additional strength, lasers and force fields. Future movie versions of the character use a different take on the character, inspired by the Ultimate version in which Doom gets powers from the same accident that caused the Four to get theirs. Doom is arguably the greatest villain in the Marvel Universe, ruling over the country Latveria, and confronting virtually every hero there is. Over time, however, he has changed his viewpoint, and since *Civil War II* believes his calling is in fact to heal the world, and is currently Iron Man, continuing Tony Stark's legacy.

*Alicia Masters:* first appeared in *Fantastic Four #8 (Nov, '62)* as Ben Grimm's blind love interest. She was also created by Lee and Kirby. Alicia's father is the Puppet Master, something that is hinted at in a future Fantastic Four film, though no film has ever used the character, and she has on occasion been involved with Johnny Storm. Since the disbanding of the Four, and following the *Secret Invasion* by the Skrulls, she has become part of a support group, helping those who replaced by Skrulls.

**Ratings:** *IMDB:* 39%

**Review:** There are a lot of problems with this film. The casting is mostly miss, with a couple of hits, while the limited budget emphasises how superhero movies really cannot be made on the cheap. This is the epitome of a B-movie; in essence a standard Roger Corman film. The story is also mostly terrible, and yet there are flashes of something better. The script is mostly good, and there are some nice ideas. Additionally (and depressingly) this is probably the best version of Doctor Doom to hit the big screen. Largely terrible, but still worth a watch.

## BLADE
### *(21/8/1998)*

A man is taken to a party which turns out to be a blood orgy for Deacon Frost's vampires; though it is broken up by the arrival of Blade who kills the majority of the vampires. As Blade tracks down Frost's right hand man – Quinn – he meets Dr Karen Jenson who was bitten by Quinn, and who hopes to find a cure to becoming a vampire, though in the process discovers something that can destroy the creatures. Frost takes control of the vampire elders with a plan to turn all vampires into the type of creature Blade himself is.

**Cast:** *Wesley Snipes (Blade), Stephen Dorff (Deacon Frost),* Kris Kristofferson (Whistler), N'Bushe Wright (Karen), Donal Logue (Quinn), Udo Keir (Dragonetti), Sanaa Lathan (Vanessa), Arly Jover (Mercury), Kevin Patrick Walls (Krieger), Tim Guinee (Curtis Webb), Traci Lords (Raquel), Eric Edwards (Pearl)

**Dir:** Stephen Norrington; **Writer:** David S Goyer; **Prod:** Peter Frankfurt, Wesley Snipes, Robert Engelman; **Music:** Mark Isham; **Exec.Prod:** Joseph Calamari, Lynn Harris, Avi Arad, Stan Lee; **DOP:** Theo Van De Sande; **Prod.Des.:** Kirk M Petruccelli; **Ed:** Paul Rubell; **Costumes:** Sanja Milkovic Hays; **New Line Cinema/Amen Ra Films/Imaginary Forces/Marvel Enterprises**; 120; $45m; $131m

**Stan Spotting:** Not on screen, sadly, but there was a deleted scene which featured Lee as a cop that discovers Quinn's smoking body.

**Notes:** The first edit of the film was longer and had a somewhat different ending, however the reception to this was quite negative and changes were made to improve it. The film was originally planned for 1992, with LL Cool J expected to take the lead, but plans were put on hold, and ultimately it was

New Line that picked the film up. The character of Whistler – created for the 1992 film outline – so impressed Marvel that they used him in **Spider-Man: The Animated Series**, meaning he appeared on television before making his big screen debut.

**Comic Notes:** *Blade:* or Eric Brooks, first appeared in *The Tomb of Dracula #10 (Jul, '73)* and was created by Marv Wolfman and Gene Colan. Initially a one off character, Wolfman used him more and more in other places, and the character became more prominent. In the comics he is English, the son of Tara Brooks, a prostitute. The movie follows the idea that Deacon Frost did indeed feed off Tara during pregnancy creating the quasi-vampire that was Eric. Jamal Afari, a jazz trumpeter, helped teach Eric how to fight vampires, and after developing a talent with a knife, he became Blade. He would make enemies of Dracula, Frost and Morbius, before becoming part of various superhero teams. Whistler's attack by vampires is inspired by a similar attack on Afari, though in the latter case, Blade had to mercy kill his mentor.

*Deacon Frost:* also debuted in *The Tomb of Dracula*, but with issue #13 (Oct, '73), and was also created by Wolfman and Colan. Frost was a scientist trying to find immortality, and on attempting to inject a young woman with vampire blood, got into a fight with the woman's boyfriend and ended up injecting himself. Frost, rather strangely, on turning someone, would generate a doppelganger of that person, and he hoped to use this skill to replace Dracula. Frost, as mentioned, fed on Tara Brooks while she was pregnant, and also turned Hannibal King. More on him later...

**Ratings:** *IMDB:* 71%; *Rotten Tomatoes:* 54%; *Metacritic:* 45

**Review:** The first of the modern Marvel movies, the film establishes pretty clearly that superhero movies can work, even if they aren't **Superman** or **Batman**. Opting for the rollercoaster action flick, what raises **Blade** up is the perfect performance by Wesley Snipes as the title character. He is actually so good you can overlook the OOT performance by Stephen Dorff. The ending is a little bit of a fumble – though much better than what was there originally – but that's a minor quibble for an otherwise great film.

# X-MEN
### *(14/7/2000)*

Senator Robert Kelly's attempts to get the "Mutant Registration Act" passed – an act to register all humans with special abilities – see opposition from mutants in two forms: the peaceful Professor Charles Xavier, and his school for gifted youngsters, which actually hides the X-Men; and the antagonistic Erik Lehnsherr and his Brotherhood of Mutants who are prepared to use deadly force to stop Kelly. While Xavier welcomes two new recruits in the form of the mysterious Logan and untouchable Marie, Lehnsherr sees an alternative use for both mutants in his struggle.

**Cast:** *Hugh Jackman (Logan/Wolverine), Patrick Stewart (Professor Charles Xavier), Ian McKellan (Erik Lehnsherr/Magneto), Famke Janssen (Jean Grey), James Marsden (Scott Summers/Cyclops), Halle Berry (Ororo Munroe/Storm), Anna Paquin (Rogue), Tyler Mane (Sabretooth), Ray Park (Toad), Rebecca Romijn-Stamos (Mystique), Bruce Davison (Senator Kelly), Matthew Sharp (Henry Gyrich), Sumela Kay (Kitty Pryde/Shadowcat), Shawn Ashmore (Bobby Drake/Iceman), Katrina Florece (Jubilation Lee/Jubilee), Alexander Burton (John Allerdyce/Pyro)*

**Dir/Writer:** Bryan Singer; **Writers:** David Hayter, Tom DeSanto; **Prod:** Lauren Schuler Donner, Ralph Winter; **Music:** Michael Kamen; **Exec.Prod:** Avi Arad, Richard Donner, Tom DeSanto, Stan Lee; **DOP:** Newton Thomas Sigel; **Prod.Des.:** John Myhre; **Ed:** Steven Rosenblum, Kevin Stitt, John Wright; **Costumes:** Louise Mingenbach; **20th Century Fox/Marvel Enterprises/ Donners' Company/Bad Hat Harry Productions**; 104; $75m; $296m

**Stan Spotting:** A hot dog vendor, of all things.

**Notes:** Originally other X-Men were planned to appear in the film, including Beast and Gambit. However, the former was too difficult to realise on the budget, and the latter wouldn't have worked as a cameo in the same way Shadowcat and Jubilee did. When the script was written, Mel Gibson was envisioned as Wolverine, and as such, the character become the driving force behind the film. Originally Dougray Scott was cast as Wolverine, but filming on **Mission: Impossible II** overran, and so the part was offered instead to Jackman. A surprisingly varied number of people auditioned for roles, including Mariah Carey and Janet Jackson for Storm, Shaquille O'Neil for Bishop (who ultimately didn't appear), Viggo Mortenson for Wolverine and Terence Stamp and Michael Jackson for Professor Xavier. Yes, you read that correctly. Angela Bassett and Rachel Leigh Cook were offered Storm and Rogue, but both refused for monetary reasons. David Hayter and Tom DeSanto both make cameo appearances in the film as cops.

**Comic Notes:** *Professor Charles Xavier, Scott Summers, Jean Grey, Bobby Drake and Magneto*: all made their first appearance in *The X-Men #1 (Sep,*

*'63)* and were created by Stan Lee and Jack Kirby. In truth, the characters are all broadly the same, but over 50 years of history have created some changes in the characters which obviously aren't addressed in this film – or indeed the film series in general. Xavier was indeed born to a wealthy family (as would be seen in **First Class**), and at a young age, discovered his telepathic abilities (like his movie counterpart), and went bald (unlike his movie counterpart). He studied at Oxford – becoming a geneticist and psychologist – and met a woman named Moira Kinross, with whom he fell in love. She would call the engagement off during the Korean War, but he would go onto meet a number of people who would influence his life, and fight an alien named Lucifer, who crippled him. Later he formed the School for Gifted Youngsters – again coming into contact with people he had met earlier on – and from there would form the X-Men. Max Eisenhardt, who would variously use the pseudonyms Magnus and Erik Lehnsherr (and also Magneto), was a holocaust survivor, who was naturally deeply affected by what happened such that when he discovered his mutant ability to manipulate magnetic fields as a youngster, he vowed to not let happen to mutants what the Nazis had done to the Jews. Becoming Magneto, he formed the Brotherhood of Evil Mutants (though one might argue this was an unwise title for his group) to fight the cause of homo superior. The X-Men, as such, were created to specifically fight against them, and protect humans. Xavier passionately believed that humans and mutants could exist in harmony, something which Eisenhardt completely disagreed with. Throughout the comics series, both men would die, be resurrected, lead the X-Men and discover they had children. It's been a journey for them. Jean Grey was a mutant whose father was friends with Charles Xavier. She had the power of telekinesis, and it was later revealed telepathy, and was sent to Xavier's School in an effort to help her. She became a member of the X-Men known as Marvel Girl. During the 1970's, Chris Claremont decided to alter the character and make her enormously powerful, changing her codename to Phoenix and giving her immense powers. They would get out of control and she would sacrifice her life to save everyone's. Latterly this was retconned as well, to reveal she had been possessed by the Phoenix Force which was what died, and the real Jean Grey was found by the Fantastic Four. She would re-join the original X-Men line up (Marvel Girl, Cyclops, Iceman, Beast, Angel) to form X-Factor. The story continues, though at the time of writing Jean is dead. Sort of (the original X-Men have been pulled through time...shall we just say it's complicated, and leave it at that?). Scott Summers is the orphan son of Christopher Summers, who has the ability to fire beams from his eyes that can only be stopped by ruby quartz. Originally he had a head injury from escaping the plane crash that killed his parents which caused him to have no control over his powers, though this was later retconned to be a mental block because

of the trauma in his life. Scott would be the first member of Xavier's school, the first X-Men and their leader for most of the time. He fell in love with Jean, and the two would remain a couple, again for most of the time, though Wolverine was an occasional problem in this relationship, and when Jean died the first time, Scott ended up marrying a woman who was almost identical to Jean – Madelyne Pryor (she was the clone of Jean, who was created by Nathaniel Essex, or Mr Sinister, and who would give birth to Nathan Summers, which Sinister needed to...well, let's just say it's complicated and leave it at that?). Bobby Drake's powers manifested when he was on a date, and this landed him in protective custody until he was rescued by Xavier. The second X-Men, he was also the youngest, and while he does indeed have control over ice, his powers have changed slightly – originally his ice form was like a snowman. He would leave the team, but re-join his old friends when they formed X-Factor.

*Ororo Munroe*: first appeared in *Giant Size X-Men #1 (May, '75)*, created by Len Wein and Dave Cockrum, as part of an effective relaunch for the X-Men. A Kenyan who was raised in Harlem, her parents moved to Cairo but were killed. Ororo became a thief, where she encountered the Shadow King (a powerful mutant) and Charles Xavier. She found herself wandering through Africa when her powers manifested and she was treated as a goddess, until Xavier approached her and recruited her. Storm would become leader of the X-Men for a time, as well as change her appearance radically after meeting Wolverine's friend Yukio, and even have her powers taken from her by Henry Gyrich (though she would get them back). For a time she was also married to T'Challa (the Black Panther) and as such was Queen of Wakanda, though the pair have since divorced.

*Wolverine*: was another in the relaunch of the X-Men, though he first appeared in *The Incredible Hulk #180 (Oct, '74)* and was created by Roy Thomas, Wein, John Romita Sr and Herb Trimpe (Thomas asked Wein to create a short Canadian named Wolverine, which Romita designed the look of, including the claws, and Trimpe drew the issue he first appeared in). James Howlett's abilities are much like the comic version, and indeed it would be retconned in the comics to show he once had bone claws which were then lined with adamantium by the Weapon X program. He adopted the name Logan after killing his biological father Thomas Logan (the Howlett's groundskeeper), and fled, joining the Canadian military, and then settling in Japan where he married a woman named Itsu and had a son named Daken (although he was unaware of this). During World War II he would be one of Captain America's soldiers, and would then join the CIA and be recruited to Team X. Returning to Canada, he became part of the Weapon X program, and was then later sent to assassinate Charles Xavier – though his memories of this were wiped by Xavier. Wolverine has made his way around the Marvel universe in various capacities, but currently is dead; a virus having stopped his healing abilities. Strangely, Spider-Man, Captain America and Deadpool have all worked for Wolverine posthumously, and Deadpool currently has the resources to resurrect Wolverine should he choose.

*Rogue:* though the movies have named her Marie D'Ancanto, Rogue has only given her first name in the comics; Anna Marie. First appearing in *Avengers Annual #10 (Nov, '81)*, she was created by Chris Claremont and Michael Golden. Similar to the movie, Rogue's powers manifested when she was kissed by her boyfriend, and she was taken by Mystique and recruited to the Brother of Mutants. Working for the Brotherhood, she attacked Ms Marvel and absorbed her powers, but the nature of Ms Marvel's personality meant that the absorption was permanent, and while she always has some psychic holdover from the people she touches, Carol Danvers' full personality lay in Rogue's mind, and she needed Xavier's help to deal with it. She became a member of the X-Men, though was untrusted by them until she willingly took a laser blast for Wolverine's fiancé Mariko, and then absorbed Colossus' powers so he could be healed. She would later lose Ms Marvel's powers when she (surprise, surprise) died, and would start an on-again-off-again relationship with Gambit, while also returning to the X-Men. She was later killed again, but has since been resurrected.

*Mystique*: first appeared in *Ms Marvel #16 (May, 78)*, though this was more a cameo, and her first full appearance was the following issue. She was created by Dave Cockrum and Chris Claremont, and has the ability to mimic the appearance and voice of anyone. Her appearance in the comics is quite different to the movie – she does have blue skin, red hair and yellow eyes, but she doesn't have any sort of scaling, and always wears clothing. Raven Darkholme's age is unknown but she met her long term lover Irene Adler around 1900. Despite her relationship with Adler, Darkholme has been sexually involved with a number of people, including Sabretooth with whom she had a son, Graydon Creed, and also Baron Christian Wagner, with whom she also had a son. She would adopt Rogue (as mentioned above) and form her own Brotherhood of Mutants, ultimately turning themselves over to the government, becoming the Freedom Force in exchange for full pardons. She would go onto to act for the government and Magneto, be a member of the X-Men as well as founding another Brotherhood.

*Victor Creed*: or Sabretooth, was a vicious young man who was abused by his father, whom he subsequently killed. He first appeared in *Iron Fist #44 (Aug, 77)*, created by Claremont and John Byrne. He joined forces with Wolverine for a time, but then killed his girlfriend when she rejected him. Like Wolverine, he joined the CIA and then Team X, and was also recruited into the Weapon X program, though this seems to have had little effect on him.

Sabretooth would fight a number of Marvel heroes before joining Gambit's Marauders and then Mystique's Brotherhood. In what may come as no surprise, he has died and been resurrected, and still works closely with Mystique.

*Toad*: (real name Mortimer Toynbee) is actually one of Magneto's original Brotherhood, first appearing in *The X-Men #4 (Mar, '64)* created by Lee and Kirby. His early appearance was a pointless character who had the ability to jump, effectively a toady to Magneto. He would leave the Brotherhood, however, and Kurt Marko (Cain Marko's father) would genetically alter him, giving him a prehensile tongue along with a new appearance. Toynbee formed his own Brotherhood, as well as joining other Brotherhood's over time. Strangely, he formed a relationship with X-Man Husk (Paige Guthrie), though he would leave her because he knew he could never be good. Robert Kelly first appeared in *Uncanny X-Men #135 (Jul, '80)*, created by Claremont and Byrne, and is a pretty good match for his movie counterpart; essentially a senator with an axe to grind against mutants. Often the target of powerful mutant groups, it was perhaps inevitable that he would be assassinated.

*Henry Gyrich*: first appeared in *The Avengers #165 (Nov, '77)*, created by James Shooter and John Byrne. A member of the NSA he was instructed to keep an eye on the Avengers, and note any problems with their operations. A man named Forge once designed a weapon to neutralise mutant's abilities, and in an attempt to neutralise Rogue, he accidentally shot Storm. Underhanded, it's perhaps no surprise he currently works for Hydra.

*Katherine "Kitty" Pryde*: first appeared in *Uncanny X-Men #129 (Jan, '80)*, created by Claremont and Byrne, and would become a surprisingly popular character. A young Jewish girl, Kitty's powers to become intangible set in when she was 13, and she was invited to join both Xavier's School, and Emma Frost's rival Massachusetts Academy, though Frost scared Kitty into going to Xavier's. She would join the X-Men, initially going by the name Sprite, she would later become Shadowcat, and form close friendships with Colossus, Wolverine and Storm. She was, for some time, the youngest X-Man. After the classic "Mutant Massacre" storyline, Kitty remained permanently intangible for a time, but would go onto join Excalibur when her powers settled again. During this time she met an alien dragon named Lockheed who would become her permanent companion. Interestingly, she is currently married to Peter Quill (Star-Lord) and is First Lady of Spartax.

*St John Allerdyce*: or Pyro, first appeared in *The Uncanny X-Men #141 (Jan, '81)*, another Claremont/Byrne creation. He is an Australian mutant with the ability to control fire, who was recruited by Mystique to join her Brotherhood of Evil Mutants. He remained with her when it became Freedom Force and, ironically given Mystique formed her Brotherhood to assassinate

Senator Kelly, he died saving Senator Kelly's life from a new Brotherhood. Of course he was resurrected...

*Jubilation Lee*: first encountered the X-Men in *Uncanny X-Men #244 (May, '89)*, at a time when they were presumed dead, though actually living in Australia. Created by Claremont and Marc Silvestri, she was an orphan – her parents killed by mobsters – eking out a living in a mall. It was here she discovered she could create mini fireworks, and fascinated by the appearance of Rogue, Dazzler, Psylocke and Storm, she followed them through their spatial gateway to Australia and hid there, ultimately coming out of hiding to assist them when they were attacked by Reavers. She formed a close bond with Wolverine, and was welcomed into the X-Men, though she swapped through teams as time went on due to her young age. She had a rough time as things went on, losing her powers in the "Decimation" storyline, and then being turned into a vampire, which she currently is; along with an adopted son (yes, it's complicated).

**Ratings:** *IMDB:* 74%; *Rotten Tomatoes:* 81%; *Metacritic:* 64

**Review: Blade** may have been the movie to launch Marvel's characters, but **X-Men** was the one that nailed, exactly, how a Marvel movie should look. Getting the casting almost perfect, Bryan Singer seems to have a brilliant understanding of what the X-Men are supposed to represent and as such they come across extremely effectively in this film. There are a few problems, but it's easy to glide over them given the quality of the film.

# BLADE II
## *(22/3/2002)*

Blade – now with the assistance of Scud – retrieves Abraham Whistler and manages to cure him with Karen's antidote. Vampire lord Damaskinos sends his daughter Nyssa to meet with Blade to arrange a truce – a deadly new form of vampire has appeared and Damaskinos suggests they join forces to destroy these Reapers. However Damaskinos is hiding a deadly secret from Blade, one which gives motivation for the leader of the Reapers.

**Cast:** *Wesley Snipes (Blade),* Kris Kristofferson (Whistler), Ron Perlman (Reinhardt), Leonor Varela (Nyssa), Norman Reedus (Scud), Thomas Kretchsmann (Damaskinos), Luke Goss (Nomak), Matthew Schulze (Chupa), Danny John Jules (Asad), Donnie Yen (Snowman), Karel Roden (Karel Kounen), Marit Velle Kile (Verlaine), Darren Crawford (Lighthammer), Tony Curran (Priest), Santiago Segura (Rush)

**Dir:** Guillermo del Toro; **Writer:** David S Goyer; **Prod:** Peter Frankfurt, Patrick J Palmer, Wesley Snipes; **Music:** Marco Beltrami; **Exec.Prod:** David S Goyer, Lynn Harris, Toby Emmerich, Michael de Luca, Avi Arad, Stan Lee; **DOP:** Gabriel Beristain; **Prod.Des.:** Carol Spier; **Ed:** Peter Amundson; **Costumes:** Wendy Partridge; **New Line Cinema/Amen Ra Films/Imaginary Forces/ Marvel Enterprises/Justin Pictures**; 117; $54m; $155m

**Notes:** Due to the amount of gore, in order to even get an R rating, the vampires had to have green blood. The character of Verlaine was originally intended to be the sister of Racquel from the previous movie. David Goyer wanted to use the character of Morbius as the villain, but Marvel had plans to use the character elsewhere – though those plans never came to fruition.

**Comic Notes:** Unusually, outside of Blade, there is no other influence from the comics.

**Ratings:** *IMDB:* 67%; *Rotten Tomatoes:* 57%; *Metacritic:* 52

**Review:** Del Toro's influence on the film is immediately felt, and this film feels, stylistically, a step up from its predecessor. The movie has an unusually varied cast, including the obligatory casting of Ron Perlman, and the story is well told, giving Blade a lot to do. The Reapers in particular come across as very disturbing. A rare case of the sequel outdoing the original.

## SPIDER-MAN
### *(30/4/2002)*

On a visit to a genetics lab, Peter Parker is bitten by a radioactive spider, which gives him superhuman abilities, including the ability to shoot web, to climb walls, and a sixth sense for danger. Norman Osborn voluntarily imbibes a formula to enhance his physical attributes, but it also drives him insane. After Peter's uncle is killed thanks to Peter's inactivity, Peter decides to fight crime as Spider-Man – and his first opponent is Osborn, now with the assumed identity of the Green Goblin.

**Cast:** *Tobey Maguire (Spider-Man/Peter Parker), Willem Dafoe (Green Goblin/ Norman Osborn), Kirsten Dunst (Mary Jane Watson), James Franco (Harry Osborn), Cliff Robertson (Ben Parker), Rosemary Harris (May Parker), J K Simmons (J Jonah Jameson),* Gerry Becker (Maximilian Fargas), *Bill Nunn (Joseph "Robbie" Robertson),* Jack Betts (Henry Balkan), Stanley Anderson

(General Slocum), *Ron Perkins (Dr Mendell Stromm), Joe Manganiello (Flash Thompson), Elizabeth Banks (Betty Brant), Michael Papajohn (Carjacker)*

**Dir:** Sam Raimi; **Writers:** David Koepp; **Prod:** Laura Ziskin, Ian Bryce; **Music:** Danny Elfman; **Exec.Prod:** Avi Arad, Stan Lee; **DOP:** Don Burgess ASC; **Prod.Des.:** Neil Spisak; **Ed:** Bob Murawski, Arthur Coburn ACE; **Costumes:** James Acheson; **Columbia Pictures/Marvel Enterprises**; 121; $139m; $825

**Notes:** This is the first film to have Marvel's flip-page logo on it, and the first film ever to make over $100 million on its opening weekend. The development of the film was long and complicated, with James Cameron attached to it for some period of time as the movie changed from company to company. Ultimately Cameron lost his chance and when Columbia acquired the rights, again a number of directors were considered before comic book fan Raimi was given the job. The theme song for the sixties *Spider-Man* cartoon appears during the movie, and also closes the film; though the theatrical version uses Aerosmith's cover, while the DVD release used the original television theme. The 2001 terrorist attack caused some changes to the movie; all posters had the twin Towers removed from them, or from the reflection in Spider-Man's mask (though they were not removed from the movie), while a new scene was added where New Yorkers rallied around Spider-Man to throw garbage at Green Goblin. Being a Sam Raimi film, both Ted Raimi and Bruce Campbell briefly appear.

**Stan Spotting:** Not as J Jonah Jameson, much to Stan's disappointment, as that was a role he was keen to play. Instead he saves a girl from some falling building. A deleted scene saw him trying to sell Peter Parker some X-Men glasses.

**Comic Notes:** *Peter Parker:* Stan Lee and Steve Ditko created Spider-Man for *Amazing Fantasy #15 (Aug, '62)*, making him one of Marvel's first characters from their post-war relaunch. He was, and currently is, the most successful Marvel superhero ever (indeed, the most successful superhero full stop!). The story of his origin is virtually identical to the comic book version (and known to almost everyone now), though in the comics Peter builds web shooters rather than is able to biologically produce it. He has an extensive history, obviously, including marrying Mary Jane, and then having that rewritten by, effectively, the devil (indeed the same one that created Ghost Rider). He has fought copious villains, been an Avenger and a member of the Fantastic Four, and is essentially the very lynchpin of the Marvel Comic universe.

*Green Goblin*: Norman Osborn first appeared in *The Amazing Spider-Man #14 (Jul, '64)*, though he wouldn't appear as Osborn for another nine issues. Like Peter, he was created by Lee and Ditko, and his backstory is also quite similar to the movie. Indeed when it was told out in *The Amazing Spider-Man #37 (Jun, '66)*, Mendel Stromm appeared as Osborn's college professor who effectively designed the Goblin serum; though by that stage Osborn had had his old friend arrested for embezzlement. Stromm, incidentally, would go on to try to get revenge on Osborn and become the Robot-Master, bringing him into conflict with Spider-Man.

*Uncle Ben, Aunt May, Eugene "Flash" Thompson and the Carjacker*: also all debuted in *Amazing Fantasy #15*, and again their backstories are all fairly similar. Perhaps of interest is that at the time, the phrase "with great power comes great responsibility" was just a caption over Ben's death, but later comics retconned this to have Ben saying it. Notably, Ben Parker's death has never been reversed in the comics, and remains virtually the only important death not to have been.

*Mary Jane Watson*: debuted in *The Amazing Spider-Man #25 (Jun, '65)*, though only partially, and was indeed mentioned ten issues earlier. She was fully revealed in *#42 (Nov, '66)* where she famously said "Face it Tiger... you've hit the jackpot!" Unlike everyone else so far, Mary Jane's creators were not only Lee and Ditko, but also John Romita Sr. Essentially a rival to Gwen Stacy for Peter's affections, Peter would later marry Mary Jane and remain so for some time, until controversially Mephisto undid the marriage in exchange for May's life.

*Harry Osborn*: another Lee/Ditko creation – first appeared in *The Amazing Spider-Man #31 (Dec, '65)*, and was indeed the rather neglected son of Norman, desperate to get his father's approval. As time went on, Harry's life would have its ups and downs. The climax of the movie is lifted from *The Amazing Spider-Man #121-#122 (Jun-Jul, '73)*, though it is Gwen rather than Mary Jane that is thrown from the bridge. In the comics Harry would discover that Peter was Spider-Man, and this would prompt him to taken on the Green Goblin mantle for a time. He would marry Liz Allen and have a son – Normie – with her.

*The Daily Bugle:* J Jonah Jameson made his first appearance in *The Amazing Spider-Man #1 (Mar, 63)*, while Elizabeth "Betty" Brant would appear three issues later in *#4 (Sep, '63)* – both by Lee and Ditko – and Joseph Robertson, or Robbie, would appear in *#51 (Aug, '67)*, though this was by Lee and Romita Sr. The three members of the Daily Bugle are very similar to their comic book counterparts. Jameson, after some small heart attacks, handed the job of editor-in-chief over to Robbie. Betty was briefly dating Ned Leeds, who would turn out to be one of the Hobgoblins, but the discovery of this would send Betty over the edge.

**Ratings:** *IMDB:* 73%; *Rotten Tomatoes:* 89%; *Metacritic:* 73

**Review:** If **Blade** proved that Marvel properties were rife for the movie making, **Spider-Man** proved just how perfect a comic book movie could be. Maguire never quite works as Peter and Kirsten Dunst seems a bit forced, but every other piece of casting in the film is a tour de force, from the brilliance of Dafoe's split personality to the perfection of Rosemary Harris as May Parker; indeed she looks like the comic character brought to life. Though, in truth, it is J K Simmons who steals the movie with a piece of casting that is unlikely to ever be bettered. The plot is a roller coaster, and the relationship between the characters makes it even more sublime. A truly superb movie.

## And the Oscar goes to…

**The Dark Knight** is, of course, the comic book movie that spectacularly won an Academy Award – though whether this was influenced by Heath Ledger's passing remains a matter for debate. Comic book movies are generally dismissed by the Academy, but often the production side of things is recognised. As such, here's a roundup of Academy Award nominations – and the occasional win – that Marvel films have had outside the MCU.

**Spider-Man:** Nominated for Best Sound Mixing (Kevin O'Connell, Greg P Russell and Ed Novick) and Best Visual Effects (John Dykstra, Scott Stokdyk, Anthony LaMolinara and John Frazier). Sadly no wins.

**Spider-Man 2**: Nominated again for Best Sound Mixing (Kevin O'Connell, Greg P Russell, Jeffrey J Haboush and Joseph Geisinger) and Best Visual Effects (Dykstra, Stokdyk, LaMolinara and Frazier), and also for Best Sound Editing (Paul N J Ottosson). Happily, Dykstra's team walked away with the Oscar.

**X-Men: Days of Future Past**: Fox getting its nomination for Best Visual Effects (Richard Stammers, Lou Pecora, Tim Crosbie and Cameron Waldbauer), but failing to take the Oscar home.

**Logan** looked like it was in with a good chance to take home an Oscar, but sadly ended up only being nominated for Best Adapted Screenplay (Scott Frank, James Mangold and Michael Green) for which it did not win.

**Spider-Man: Into the Spider-Verse** actually won the Oscar for Best Animated Feature, but as an animated feature (criminally) nobody counts it.

The Marvel Cinematic Universe has been nominated a number of times, and took home a slew of Oscars for **Black Panther**, though all were practical.

## DAREDEVIL
### *(14/2/2003)*

Blind lawyer Matt Murdock spends his nights as a costumed vigilante thanks to a chemical spill which granted him superhuman senses. When businessman Nikolas Natchios welches on a deal with city crime baron Wilson Fisk, Fisk sends Bullseye to kill him, and Murdock – as Daredevil – fails to stop him. Matt tries to locate Bullseye, but finds himself up against Elektra, Natchios' daughter, who blames him for her father's death.

**Cast:** *Ben Affleck (Matt Murdock/Daredevil), Jennifer Garner (Elektra Natchios), Michael Clarke Duncan (Wilson Fisk/The Kingpin) and Colin Farrell as Bullseye, Joe Pantoliano (Ben Urich), Jon Favreau (Franklin "Foggy" Nelson), David Keith (Jack Murdock),* Erick Avari (Nikolas Natchios), Paul Ben-Victor (Jose Quesada), Derrick O'Connor (Father Everett), *Leland Orser (Wesley Owen Welch),* Scott Terra (Young Matt), *Ellen Pompeo (Karen Page),* Kevin Smith (Jack Kirby, Forensic Assistant), Stan Lee (Old Man at Crossing), Frank Miller (Man with Pen in Head)

**Dir/Writer:** Mark Steven Johnson; **Prod:** Arnon Milchan, Gary Foster, Avi Arad; **Music:** Graeme Revell; **Exec.Prod:** Stan Lee, Bernie Williams; **DOP:** Ericson Core; **Prod.Des.:** Barry Chusid; **Ed:** Dennis Virkler ACE, Armen Minasian; **Costumes:** James Acheson; **20th Century Fox/Regency Enterprises/New Regency/Horseshoe Bay/Marvel Enterprises**; 103; $78m; $179m

**Stan Spotting:** Shockingly, the old man at the crossing that Matt Murdock stops from being hit by a car.

**Notes:** A number of characters are named for Marvel staff associated with Daredevil: Father Everett, for Bill Everett, one of Daredevil's creators; Jose Quesada for Joe Quesada, Daredevil artist and former Marvel editor in chief; Colan for Gene Colan, John Romita for both John Romita Sr and his son, Kane for Gil Kane, Miller and Mack for Frank Miller and David Mack and Jack Kirby for Jack Kirby, all of whom were artists on the comic at some point.

Bendis is named for Daredevil writer Brian Michael Bendis. Additionally, writers Stan Lee, Frank Miller and Kevin Smith all appear in the movie. A director's cut of the movie includes a large number of scenes as well as an entire subplot that was removed. It is widely regarded as far superior to the theatrical version.

**Comic Notes:** *Matt Murdock, Karen Page and Foggy Nelson:* all first appeared in *Daredevil #1 (Apr, '64)* and were created by Stan Lee and Bill Everett (and possibly also Jack Kirby). Murdock's backstory and powers are largely the same as the series, though the comics make it explicit that his heightened senses are a superpower derived from the radioactive waste. Nelson's story is also much the same, but Page had a much more interesting background in the comics (though given she's barely in the movie that's hardly surprising), including a rather dark storyline about being addicted to heroin and doing pornography. I wonder why that never came up in the film?

*Ben Urich:* first appeared in *Daredevil #153 (Jul, '78)*, created by Roger Mackenzie and Gene Colan. His character is broadly the same, though he plays a large part in the various *Spider-Man* comic series as well. In the comics he does indeed know the secret identity of Daredevil (and Spider-Man, curiously), though he is one of the most trusted associates of both heroes.

*Wilson Fisk:* more commonly known as the Kingpin; first appeared in *The Amazing Spider-Man #50 (Jul, '67)* and was created by Lee and John Romita Sr. Physically he differs from his movie counterpart, though both are big and bald, the comic version is Caucasian. In the comics he is also married to Vanessa and has a son named Richard (or, had...). The Kingpin is one of the "big" villains of Marvel, and has confronted most superheroes over time.

*Wesley:* is very much a background character, appearing first in *Daredevil #227 (Feb, 86)*. Created by Frank Miller and David Mazzucchelli, he has only appeared in four issues.

*Elektra Natchios:* first appeared in *Daredevil #168 (Jan, '81)*, created by Frank Miller. The daughter of Hugo and Christine, and younger sister to Orestez, she has a complicated backstory that is inevitably different to the movie, simply because the comics have given her two. Christine was killed in different ways, though in both she gave birth to Elektra just before she died. She was assaulted by kidnappers, but rescued by Orestez who suggested she get martial arts training. Another account suggests she was possibly molested by her father, who then sent her away for therapy. Ultimately, at the age of 19, she went to America where her father was made Greek Ambassador, and at school she dated Matt Murdock. When her father was killed by terrorists, she joined the Hand, who trained her as an assassin, and it was on a mission that she later encountered Daredevil. Still later she was killed by Bullseye while working for the Kingpin. Stone resurrected her, though more on that to com...

It's worth noting that Frank Miller only intended Elektra to appear in a single issue, and Marvel's reuse of the character upset him as he had been promised that they wouldn't do so.

Bullseye: first appeared in *Daredevil #131 (Mar, '76)* and was created by Marv Wolfman and John Romita Sr. Named Lester Poindexter here, in the comics he's used both Lester and Benjamin Poindexter as real names. Nothing is known about his past in the comics, outside of the fact his father was abusive and he was placed in a foster home. His perfect aim allowed him to be a brilliant baseballer, but this came to an end when he killed another player by throwing a bat at his head. He has remained one of Daredevil's most prominent villains, notably killing Karen Page. At one point Norman Osborn made him Hawkeye in the Dark Avengers.

**Ratings:** *IMDB:* 53%; *Rotten Tomatoes:* 44%; *Metacritic:* 42

**Review:** Sadly, a movie that seems intent on aping Tim Burton's **Batman**, even down to mimicking some of the scenes from that film and Revell doing a subpar Danny Elfman. Affleck isn't terrible, though Farrell is, and Garner is badly miscast. In fact, it's Michael Clarke Duncan who steals the movie, giving a bravura performance as the Kingpin, but with a wandering plotline and uneven tone, the film stumbles more often than it runs.

## X2
### *(24/4/2003)*

The attack on the President by a mutant with teleportation, prompts the President to allow William Stryker and his cover strike team to attack Xavier's school, based on information he has obtained from Magneto. Mystique frees Magneto, while Wolverine goes on the run with some of the kids from the school and Storm and Jean Grey find the teleporting mutant. With Stryker having kidnapped Xavier and Cyclops, the X-Men join forces with Magneto in an attempt to stop him.

**Cast:** *Patrick Stewart (Professor Charles Xavier), Hugh Jackman (Logan/ Wolverine), Ian McKellan (Erik Lehnsherr/Magneto), Halle Berry (Ororo Munroe/Storm), Famke Janssen (Jean Grey), James Marsden (Scott Summers/Cyclops), Anna Paquin (Rogue), Rebecca Romijn-Stamos (Mystique), Brian Cox (William Stryker), Alan Cumming (Kurt Wagner/ Nightcrawler), Bruce Davison (Senator Kelly), Aaron Stanford (John Allerdyce/ Pyro), Shawn Ashmore (Bobby Drake/Iceman), Kelly Hu (Yuriko Oyama/Lady Deathstrike), Katie Stuart (Kitty Pryde/Shadowcat), Kea Wong (Jubilation Lee/*

*Jubilee), Bryce Hodgson (Artie), Shauna Kain (Theresa Cassidy/Siryn), Daniel Cudmore (Piotr Rasputin/Colossus), Steve Bacic (Dr Hank McCoy), Michael Reid Mackay (Jason 143)*

**Dir/Writer:** Bryan Singer; **Writers:** David Hayter, Michael Dougherty, Dan Harris, Zak Penn; **Prod:** Lauren Schuler Donner, Ralph Winter; **Music:** John Ottman; **Exec.Prod:** Avi Arad, Bryan Singer, Tom DeSanto, Stan Lee; **DOP:** Newton Thomas Sigel; **Prod.Des.:** Guy Hendrix Dyas; **Ed:** John Ottman, Elliot Graham; **Costumes:** Louise Mingenbach; **20th Century Fox/Marvel Enterprises/Donners' Company/Bad Hat Harry Productions**; 134; c $120m; $407

**Stan Spotting:** Not this time. But Stan Lee will return!

**Notes:** Although the title of the movie as it appears on screen is **X2**, some parts of the world advertised the film as **X-Men 2**, or **X2: X-Men United**. There are an enormous number of references to the X-Men comics, including a massive list of characters on Lady Deathstrike's computer. Michael Dougherty and Dan Harris cameo as Weapon X surgeons, while Bryan Singer cameos as a security guard to Magneto. John Ottman's fantastic X-Men Theme would be used in later films he scored.

**Comic Notes:** *William Stryker:* first appeared in *X-Men: God Loves, Man Kills (Jan, '83)*, created by Chris Claremont and Brent Anderson. This movie is heavily inspired by that graphic novel as well. There is a slight difference between the comic and movie versions of the character. In the comics, Stryker is no longer part of the military, and it's not entirely clear if he was involved in the Weapon X project. He is actually a religious fanatic; a preacher determined to end mutants, even going so far as to kill his wife when she gave birth to a mutant, Jason, whom he also stabbed, before blowing then all up. William survived the explosion to start a campaign of genocide, but Jason also survived and had his mutant abilities removed by AIM. William was arrested, but was later rescued by his lover, Lady Deathstrike, and he then launched a new campaign against the X-Men, particularly Shadowcat, whom he held a grudge against. Kitty, however, managed to reach him, and got him to see the error of his ways. This didn't last long, sadly. Jason, meanwhile, grew up and continued his father's work with a group called the Purifiers.

  *Yuriko Oyama:* first appeared in *Daredevil #197 (Aug, '83)* and was created by Dennis O'Neil and Larry Hama. She was the daughter of a Japanese crime boss who discovered the process of bonding adamantium to bone, and who scarred her and her brothers for their shame. She initially fought alongside Daredevil against her father, but her lover was also in his service and he committed suicide when Oyama was killed. She attempted to avenge her father, but ended up being transported to the Mojoverse, where she had her body cybernetically upgraded. As Lady Deathstrike she first appeared in *Alpha Flight #33 (Apr, '86)*, and this new variation of the character is attributed to writers Bill Mantlo and Claremont, and artists Sal Buscema and Barry Windsor-Smith. She joined the Reavers, as well as becoming William Stryker's lover. Recently she died, but her consciousness remains intact inside the body of her friend Reiko.

*Kurt Wagner and Peter Rasputin*: first appeared in the relaunch *Giant Size X-Men #1 (May, '75)*, created by Len Wein and Dave Cockrum as part of the new X-Men team. Like his movie counterpart, Wagner has the ability to teleport, a gift he got from his father Azazel. His mother was Raven "Mystique" Darkholme, who abandoned him when a mob found out about them, though Azazel rescued him and left him with Margali Szardos, who then gave him to a Bavarian circus. In the comics, Nightcrawler is also blue with a tail, though he is actually furry, and his hands and feet have noticeably unique differences. When Kurt tried to stop one of the circus boys from killing a local villager, the villagers turned on him, but he was rescued by Xavier and joined the X-Men. During the "Mutant Massacre" he was badly injured, and when he recovered, along with Kitty Pryde, would join Excalibur, though they later re-joined the X-Men. Naturally, Nightcrawler died in battle, and has since been resurrected. In the comics, Peter Rasputin is properly named Piotr – Peter being a nickname – and he is the older brother of Illyana, and younger of Mikhail. Born in Serbia, his power manifested when he rescued Illyana from a runaway tractor. After this he was recruited to join the X-Men. He was probably the most pacifistic of the team, though during the "Mutant Massacre", his rage got the better of him when Kitty – with whom he was in love – was harpooned, and he killed Riptide. He was also attacked, and his body was locked in his metal form, though he re-joined the X-Men and during *The Fall of the Mutants*, he left with them for Australia to remain hidden there. Over time he has been involved in a number of incidents regarding his siblings, but the death of Illyana forced him to leave the X-Men and join the more radical Acolytes. In a surprise first for the X-Men, Colossus was killed and then later resurrected. He currently serves on the X-Men team led by Wolverine.

*Theresa Cassidy*: is the daughter of Sean Cassidy in the comics (though the movie timeline doesn't automatically exclude this possibility), and upon learning of her sonic powers, adopted the name Siryn. First appearing in *Spider-Woman #37 (Apr, '81)*, she was created by Claremont and Steve Leialoha. Theresa's mother Maeve was killed in an IRA bombing, which Sean blamed Black Tom Cassidy for, and Black Tom got his revenge by raising

Theresa but not telling Sean of her existence. Spider-Woman fought and beat Black Tom in one of his criminal enterprises, and handed Theresa over to the X-Men, where she was reunited with Sean. Never an X-Man, Theresa remained on the periphery of the team, but would later give birth and name her child Sean; her father having passed away. Similarly she adopted the name Banshee in his honour, and went on to lead X-Factor. Surprisingly she hasn't died or been resurrected (though Sean has done both, of course).

Artie: the boy with the blue forked tongue is named Artie, and as Artie Maddicks appears on Lady Deathstrike's computer, we can assume that this is the character he is supposed to be. In the comics he is radically different. First appearing in *X-Factor #2 (Mar, '86 – Volume 1)*, he was created by Bob Layton and Jackson Guice. He looks rather unique – quite pink, with an oversized head with circular raised lumps. He became a ward of X-Factor after his father died, and it turned out he had the ability to project images, though he was mute. Mostly a bit player, he was good friends with Leech, but has since been de-powered, and has mostly disappeared.

**Ratings:** *IMDB:* 75%; *Rotten Tomatoes:* 86%; *Metacritic:* 68

**Review:** Singer amps up his movies a gear here, with a film that is even tighter than its predecessor, despite the addition of several more mutants. The Stryker storyline is played out very well, and all of this leads up to a moving ending with a hint of things to come. Singer's brilliance cannot be understated in regards to this film, which is genuinely fantastic.

## HULK
### *(20/6/2003)*

David Banner attempted to create super soldiers using genetic modification, but his experiments were stopped by General Ross before testing them on humans; though Banner tested on himself. However, his son Bruce inherited some of the genetic mutation, and during experiments conducted by Bruce and his ex-girlfriend Betty Ross (the General's daughter), a burst of gamma radiation hits Bruce and triggers a transformation that is unstable, and brought on by anger. While Major Talbot wants to use Banner as a weapon, David Banner tries to recreate Bruce's experiment and triggers his own transformation, and General Ross becomes convinced that Banner needs to be executed as a serious threat.

**Cast:** *Eric Bana (Bruce Banner), Jennifer Connelly (Betty Ross), Sam Elliott (Ross), Josh Lucas (Talbot), Nick Nolte (Father),* Cara Buono (Edith Banner),

Paul Kersey (Young David Banner), Kevin Rankin (Harper), Celia Weston (Mrs Krensler)

**Dir:** Ang Lee; **Writers:** James Schamus, John Turman, Michael France; **Prod:** Avi Arad, Gale Anne Hurd, Larry J Franco, James Schamus; **Music:** Danny Elfman; **Exec.Prod:** Kevin Feige, Stan Lee; **DOP:** Frederick Elmes; **Prod.Des.:** Rick Heinrichs; **Ed:** Tim Squyres; **Costumes:** Marit Allen; **Universal Pictures/Marvel Enterprises/Valhalla Motion Pictures/Good Machine**; 138; $137m; $245m

**Notes:** General Ross is never referred to by his nickname, "Thunderbolt", but his helicopter is called the T-Bolt. The Marvel logo is changed to green for this movie. When Marvel Studios made **The Incredible Hulk**, Sam Elliott hoped to reprise the role of General Ross.

**Stan Spotting:** One of the security guards at the facility.

**Comic Notes:** *Bruce Banner:* and his hulkish alter ego first appeared in *The Incredible Hulk #1 (May, '62)*, and like virtually all Marvel characters in the sixties, was created by Stan Lee and Jack Kirby. The Hulk was initially grey, but later became green for publishing reasons (Stan Goldberg's grey in the first issue varied so Lee chose to change it to green). In the comics he was affected by gamma radiation which meant that as he got angry, he would lose control and become the Hulk. Over the years, the Hulk has actually undergone a variety of transformations, including returning to being grey (with a resultant increase in intelligence), and being dispatched from Earth by the so-called Illuminati, who feared the danger he would bring to the planet. Most recently, Bruce Banner was killed by Clint Barton, at Banner's request to stop him from becoming the Hulk. Amadeus Cho extracted Banner's abilities and currently becomes the Hulk.

*Betty Ross and Thaddeus "Thunderbolt" Ross:* also make their first appearance in *The Incredible Hulk #1*. Thunderbolt's backstory is similar to his movie counterpart, though Betty was not a cellular biologist in the comics. Interestingly, both Ross' have become red versions of the Hulk in more recent years.

*Talbot:* first appeared in print in *Tales to Astonish #61 (Nov, '64)* created by Stan Lee and Steve Ditko. Similarly to the movie, Colonel Glenn Talbot in the comics is assigned duties by General Ross to recover and contain Bruce Banner.

*And:* David Banner is an amalgam of a variety of characters, named David in homage to the seventies television series. Bruce Banner's father was named Brian in the comics, and first appeared in *The Incredible Hulk #267 (Jan, '82)*, created by Bill Mantlo and Sal Buscema. He was abused by his father, and in turn abused Bruce – leading to many of Bruce's anger issues. It was not until the 2010's when he was given any sort of powers (effectively another version of the Hulk). Banner's movie transformation is based more on The Absorbing Man, who first appeared in *Journey into Mystery #114 (Mar, '65)*, created by Lee and Kirby. He was a former boxer who discovered he had the ability absorb the properties of whatever he touched which he gained from a magical potion.

**Ratings:** *IMDB:* 57%; *Rotten Tomatoes:* 61%; *Metacritic:* 54

**Review:** There's a lot of trying in this movie, but for the most part, not an awful lot of success. The performances are, perhaps, overthought, so that Bana occasionally comes across as less controlled than emotionless, and Connelly also lacks a strong chemistry with Bana. Lee's ideas to have the movie look like a comic book are visually jarring for the most part but all of this pales thanks to a madly misjudged performance by Nick Nolte who is totally over the top. It's not a terrible film, but the Norton version is a better pitch.

## Can Hulk Be Considered Part Of The Marvel Cinematic Universe?

The truth is, it's difficult to see how. The flashback sequences in **The Incredible Hulk** showing how the Norton Banner becomes the Hulk are completely different to the Bana Banner origin, but more than that, of course, is that they fundamentally come from two different places, both of which are established in more than just the origin sequence. Based on that alone it's difficult to see how they could be a continuation.

More importantly than that, however, is that in **The Incredible Hulk**, Betty does not know initially that Bruce is the Hulk until later. This might seem odd, but she was attacked when Bruce first changed, so perhaps that affected her memory. Either way, she clearly learnt of his identity in **Hulk**.

And yet, the ending of **Hulk** does dovetail neatly into **The Incredible Hulk**, and the timing is even neater.

In the old days of **Star Wars**, before Disney bought LucasFilm, there were a variety of levels of "canon" – some things were completely canonical, others were given levels, so that certain parts were canonical but not everything. It's tempting to do a similar thing here. It would work if we considered parts of **Hulk** canonical, insofar as the story happened, but the origin aspects were different, and for some reason Betty forgets who the Hulk actually is (or

## THE PUNISHER
### (16/4/2004)

An FBI Operation kills Bobby Saint, son of crime boss Howard Saint, and Saint orders a hit on the family of the FBI Agent who set up the sting – Frank Castle. Castle survives, though his family does not, and after his recovery, returns to Tampa where he lives in an old tenement with three unlikely neighbours. Step by step, Castle begins to take his revenge on Saint, sewing distrust and death in Saint's operation.

**Cast:** *Thomas Jane (Frank Castle),* John Travolta (Howard Saint), Will Patton (Quentin Glass), Roy Scheider (Frank Castle Snr), Laura Harring (Livia Saint), *Ben Foster (Spacker Dave), Rebecca Romijn (Joan), John Pinette (Bumpo),* Samantha Mathis (Mary Elizabeth Castle), Marcus Johns (Will Castle), Russell Andrews (Agent Jimmy Weeks), James Carpinello (Bobby & John Saint), Eddie Jemison (Mickey Duka), Eduardo Yanez (Mike Toro), Omar Avila (Joe Toro), *Kevin Nash (The Russian)*

**Dir:** Jonathan Hensleigh; **Writers:** Jonathan Hensleigh, Michael France; **Prod:** Avi Arad, Gale Anne Hurd; **Music:** Carlo Siliotto; **Exec.Prod:** Christopher Eberts, Andrew Golov, Patrick Gunn, Amir Malin, Christopher Roberts, Richard Saperstein, Andreas Schmid, John Starke, Kevin Feige, Stan Lee; **DOP:** Conrad W Hall; **Prod.Des.:** Michael Z Hanan; **Ed:** Steven Kemper, Jeff Gullo; **Costumes:** Lisa Tomczeszyn; **Marvel Enterprises/Valhalla Motion Pictures, VIP2/VIP3/Artisan**; 123; $33m; $54m

**Notes:** A longer version of the movie was on the DVD rerelease, reinstating a particularly dark subplot in which Castle discovered he was set up by Jimmy Weeks, and then drove the man to suicide. It was removed from the theatrical version, purely for pacing reasons. Originally France had the character of Microchip in the screenplay, but Hensleigh removed him because he despised the character.

**Comic Notes:** A lot of the plot is inspired by the "Welcome Back, Frank" storyline. Joan, Bumpo and Spacker Dave all first appeared in *The Punisher #1 (Apr, '00 – Vol 4)*, created by writer Garth Ennis and artist Steve Dillon. The Russian appears later in *#8 (Nov, '00)*, by the same creative team. Broadly speaking, all four characters are roughly the same, though in the comics it is the Gnucci family that is the target of Castle's attacks, and as such it is they who hire the Russian. Castle himself has his backstory altered to make him a veteran of the Gulf War in order to keep the movie current. Additionally his family is killed by the Saint family, and he is an FBI agent. The scene of the popsicle torture is lifted from *Punisher: War Zone*.

**Ratings:** *IMDB:* 65%; *Rotten Tomatoes:* 29%; *Metacritic: 33*

**Review:** A vast improvement on the previous attempt, perhaps the biggest fault in this film is Travolta who is too busy chewing the scenery to be taken as a serious villain. The plot of the film, however, is very entertaining, and Jane – whilst still not the best choice – does an acceptable job as Castle.

## SPIDER-MAN 2
### *(25/6/2004)*

Peter's personal life takes a back seat as he continues to fight crime as Spider-Man, but in doing so he doesn't notice May's financial problems, and the fact Mary Jane is now dating J Jonah Jameson's son John. Harry Osborn finances Otto Octavius and his experiments in fusion power, but a fault causes Octavius' robot arms to become fused to him. Determined to finish his experiment, Octavius demands tritium from Harry Osborn, who agrees to give it to him in exchange for the life of Spider-Man.

**Cast:** *Tobey Maguire (Spider-Man/Peter Parker), Kirsten Dunst (Mary Jane Watson), James Franco (Harry Osborn), Alfred Molina (Dr Octopus/Dr Otto Octavius), Rosemary Harris (May Parker), J K Simmons (J Jonah Jameson),* Donna Murphy (Rosalie Octavius), *Daniel Gillies (John Jameson), Dylan Baker (Dr Curt Connors), Bill Nunn (Joseph "Robbie" Robertson),* Vanessa Ferlito (Louise), Aasif Mandvi (Mr Aziz), *Willem Dafoe (Green Goblin/Norman Osborn), Cliff Robertson (Ben Parker), Elizabeth Banks (Betty Brant)*

**Dir:** Sam Raimi; **Writers:** Alfred Gough, Miles Millar, Michael Chabon, Alvin Sargent; **Prod:** Laura Ziskin, Avi Arad; **Music:** Danny Elfman; **Exec.Prod:** Stan Lee, Kevin Feige, Joseph M Caracciolo; **DOP:** Bill Pope ASC; **Prod.Des.:** Neil Spisak; **Ed:** Bob Murawski; **Costumes:** James Acheson, Gary Jones; **Columbia Pictures/Marvel Enterprises**; 127; $200m; $789m

**Notes:** Again, both Bruce Campbell and Ted Raimi make cameo appearances in the film. Dylan Baker agreed to an appearance in both this film and the next, on the understanding that when the fourth film was produced, the Lizard would be the villain. As it transpired the Lizard was indeed the villain of the fourth Spider-Man film, but unfortunately, because it was a reboot, Baker lost the part. Towards the end of the film, Mary Jane runs past someone who bears a striking resemblance to Thomas Jane, and is dressed all in black. It's unclear if this is supposed to be a nod to the Punisher, but rumour has it the man in black was Thomas Jane's stunt double, suggesting it was indeed a surprise cameo.

**Stan Spotting:** Stan's still saving people from falling buildings.

**Comic Notes:** *Otto Octavius*: first appeared in *The Amazing Spider-Man #3 (Jul, '63)*, and was another Stan Lee and Steve Ditko creation. The backstory for the comics' version is much the same as the movie version, though there are some differences – Octavius wasn't married in the comics, though he did get engaged to Mary Alice Anders, and called it off at his mother's insistence. When his mother died, Octavius become embittered and obsessed with his work. One explosion later and he was fused to the mechanical arms he used for research, and a life of crime followed. Octavius would go onto plague Spider-Man for most of his life, though interestingly, at one point his mind was placed in Peter Parker's body and he became Spider-Man.

*John Jameson*: made his debut alongside his father in *The Amazing Spider-Man #1 (Mar, '63)*, and was also an astronaut, like his movie counterpart. However the comic version found the Godstone on the moon, and when it grafted itself onto his body, John became the Man-Wolf. The transformation was only semi complete and so Man-Wolf was effectively berserk. He would later be transported to the "Other Dimension", where he would gain complete control of his transformation and call himself Stargod. He was never engaged to Mary Jane, acted as both friend and enemy to Spider-Man, and like so many others, was created by Lee and Ditko.

*Dr Curt Connors*: would appear five issues later in #6 *(Nov, '63)* – again created by Lee and Ditko – where the one armed scientist (having lost an arm during war) took a formula of reptilian DNA designed by himself and Ted Sallis (the Man-Thing) and transformed into the Lizard; a giant reptile creature that was slightly insane. The Lizard has often fought Spider-Man and other heroes (notably the X-Men), but Curt Connors remains a close friend and ally of Peter Parker's. A lot of this story would be similar to the Lizard's actual movie appearance in **The Amazing Spider-Man**.

*And:* The shot of Peter dumping the Spider-Man costume in the bin and walking away is a frame recreation from *The Amazing Spider-Man #50*, in which Peter famously gave up his superhero life for a while.

**Ratings:** *IMDB:* 73%; *Rotten Tomatoes:* 93%; *Metacritic:* 83

**Review: Spider-Man 2** is, quite simply, one of the best movies ever produced, let alone one of the best comic book movies. Everyone from the previous film returns, bringing their A-game, with J K Simmons again stealing every scene he's in, but this time you can add Alfred Molina to the mix, who delivers one of the best villain performances ever. Dr Octopus is realised fantastically well, with great sympathy. There's so much to love, be it Raimi's flawless direction, Pope's beautiful cinematography or Elfman's gorgeous score, that to list absolutely everything would take far more space than is available. The best Spider-Man movie ever made.

## BLADE: TRINITY
### *(8/12/2004)*

The original vampire, Dracula (or Drake), is resurrected by Danica Talos who hopes to use him to destroy Blade. Talos also uses her familiars to begin an FBI manhunt for Blade, which is successful when Whistler is killed. However, the Nightstalkers − a group of vampire hunters led by Hannibal King and Abigail Whistler − rescue Blade in order to stop Drake, and release the Daystar − a device that will kill all vampires.

**Cast:** *Wesley Snipes (Blade),* Kris Kristofferson (Whistler), Jessica Biel (Abigail Whistler), *Ryan Reynolds (Hannibal King), Dominic Purcell (Drake),* Parker Posey (Danica Talos), Callum Keith Rennie (Asher Talos), Triple H (Jarko Grimwood), John Michael Higgins (Dr Edgar Vance), Mark Barry (Chief Martin Vreede), Patton Oswalt (Hedges), Paul Anthony (Wolfe), Françoise Yip (Virago), Michael Anthony Rawlins (Wilson Hale), James Remar (Ray Cumberland), Natasha Lyonne (Sommerfield), Haili Page (Zoey Sommerfield)

**Dir/Writer:** David S Goyer; **Prod:** Peter Frankfurt, David S Goyer, Lynn Harris Wesley Snipes; **Music:** Ramin Djawadi, The RZA; **Exec.Prod:** Toby Emmerich, Cale Boyter, Avi Arad, Stan Lee; **DOP:** Gabriel Beristain; **Prod.Des.:** Lucrezia Casta, Chris Gorak; **Ed:** Conrad Smart, Howard E Smith; **Costumes:** Laura Jean Shannon; **New Line Cinema/Shawn Danielle Productions Ltd/Amen Ra Films/Imaginary Forces/Marvel Enterprises/ Peter Frankfurt Productions**; 113; $65m; $132m

**Notes:** A film that appears to have been a rather difficult production, as both Snipes and Kristofferson were unhappy with the direction the film was going. Snipes and Goyer had a falling out which resulted in Snipes communicating with Goyer through post-it notes signed Blade. After the movie was released, Snipes filed suit against New Line for reducing his screen time in favour of Reynolds and Biel. The suit was settled, but no details are known, and the following year, Snipes was sent to jail for tax evasion. Three endings were shot for the film. Aside from the theatrical ending, a second ending in which it was hinted Blade was succumbing to the thirst was filmed, and used on the director's cut, as it was Goyer's preferred ending. A third ending saw the Daystar being successful, but the Nightstalkers taking on werewolves.

**Comic Notes:** *Hannibal King:* first appeared in *The Tomb of Dracula #25 (Oct, '74),* created by Marv Wolfman and Gene Colan. Unlike the movies, when the comic version of King was bitten by Deacon Frost, he did not escape being turned. As such he refuses to use his vampiric abilities. He was a private detective, which he continued after his turning – but would later join forces with Blade to fight against the likes of Frost and Morbius.

*Dracula:* was, of course, created by Bram Stoker, but the Marvel version first appeared in *The Tomb Of Dracula #1 (Apr, '72)* and was created by Gerry Conway and Colan, although earlier, a version did appear in *Suspense #7 (Mar, '51)* in the story "Dracula Lives". The character is essentially the same as the Stoker version, which is not really the same as the movie version. His powers are also much the same as the original version. In the comics he crosses swords with a number of superheroes, but was killed by his son Xaras. Originally he dressed as a Victorian nobleman, but later he dressed more as an out of time medieval warrior.

**Ratings:** *IMDB:* 59%; *Rotten Tomatoes:* 25%; *Metacritic:* 38

**Review:** David S Goyer is a talented filmmaker with enough credits – particularly comic book movie credits – to give him free reign, however here it just hasn't worked. The decision to reduce Blade's screen time in favour of King and Abby was a bad one, and the film lacks Snipes' brutal charm. Equally the portrayal of Drake just doesn't work – either as a character, or as the villain. The spirited performances of Reynolds and Posey sadly can't save this movie.

# ELEKTRA
## *(14/1/2005)*

Resurrected by Stick, Elektra is trained in Kimagure, but is sent away as she is unable to contain her rage. Later she is an assassin, and one job sees her on an island awaiting her targets. She befriends Mark Miller and his daughter Abby, but then discovers they are her targets. Rather than kill them, she protects them when The Hand, led by Kirigi, come after her, and she is soon joined by Stick and his Chaste, who reveal that Abby is a martial arts prodigy that both the Chaste and the Hand seek.

**Cast:** *Jennifer Garner (Elektra),* Goran Visnjic (Mark Miller), Kirsten Prout (Abby Miller), *Will Yun Lee (Kirigi),* Cary-Hiroyuki Tagawa (Roshi) *and Terence Stamp (Stick), Natassia Malthe (Typhoid), Bob Sapp (Stone), Chris Ackerman (Tattoo),* Kurt Max Runte (Nikolas Natchios), Colin Cunningham (McCabe)

**Dir:** Rob Bowman; **Writers:** Zak Penn, Stu Zicherman, Raven Metzner; **Prod:** Arnon Milchan, Gary Foster, Avi Arad; **Music:** Christophe Beck; **Exec.Prod:** Stan Lee, Mark Steven Johnson, Brent O'Connor; **DOP:** Bill Roe ASC; **Prod.Des.:** Graeme Murray; **Ed:** Kevin Stitt ACE; **Costumes:** Lisa Tomczeszyn; **20th Century Fox/Regency Enterprises/New Regency/ Horseshoe Bay/Marvel Enterprises**; 97; c $50m; $57m

**Notes:** Jason Isaacs has an uncredited cameo as DeMarco, the man killed by Elektra in the opening sequence, and Ben Affleck reprised Daredevil in a scene that was ultimately cut from the film.

**Comic Notes:** *Stick:* first appeared in *Daredevil #176 (Nov, '81),* created by Frank Miller. His personality was very similar to how it appears here, but he led the Chaste, and sought to keep it pure. The Hand (an order of evil ninjas, created by Miller two issues earlier) were opposed to this and attempted to assassinate Stick. Stick trained both Matt Murdock and Elektra Natchios, and was later killed by the Hand, though the Chaste guard over his spirit.

    *Kirigi and the Hand:* first appeared in *Daredevil #174 (Sep, '81),* created by Frank Miller. In the comics he was resurrected by the Hand and proceeded to work for them, with the aim of killing Elektra.

    *Stone*: first appeared in *Daredevil #187 (Oct, '82),* created by Frank Miller and Klaus Janson. Stone's comic counterpart is very different to the version in the movie. He was another of Stick's pupils – his favourite in fact – and fought alongside Stick on many occasions. However, significantly, he later helped Daredevil resurrect Elektra after she was killed by Bullseye.

*Tattoo*: does have a comics' counterpart who was a female mutant who attended Xavier's School for Gifted Youngsters. She first appeared in *New X-Men #126 (Jul, '02)* and was created by Grant Morrison and Frank Quitely. She originally had the ability to form words or patterns on her skin. In truth there is very little similarity between the two versions.

*Typhoid*: better known in the comics as Typhoid Mary aka Mary Walker – first appeared in *Daredevil #254 (May, '88)* and was created by Ann Nocenti and John Romita Jr. The comic version doesn't have pestilent breath, rather she has low level psionic abilities such as telekinesis. She has a much more interesting backstory, however. A former prostitute, when Matt Murdock (before becoming Daredevil) tracked down a villain to the brothel Mary worked out, Matt accidentally pushed her out of a window, giving her brain damage. This was later retconned to her being abused as a child. She has dissociative identity order – a timid Mary personality, a violent Typhoid personality and a sadistic Bloody Mary personality. She has loved Daredevil and the Kingpin, and worked for the latter, as well as Doctor Doom. One of Daredevil's more fascinating adversaries, it's a shame she's been watered down for this movie.

*And:* the sequence with DeMarco at the beginning of the film is virtually identical to *Elektra #23 (Jul, '03 – Vol 2)*, by Robert Rodi and Sean Chen.

**Ratings:** *IMDB:* 48%; *Rotten Tomatoes:* 10%; *Metacritic:* 34

**Review:** Garner is back, and still seems wrong on every level to play Elektra. Further, after the lack of success with **Daredevil**, despite pushing ahead with this film, the movie tries to avoid any reference to **Daredevil** whatsoever, resulting in a complete lack of depth for the film. On top of that, there's a plot that is convoluted and never really fully explained, and a number of moments that are borderline ridiculous. It's trite and shallow, and a waste of a number of great characters.

## MAN-THING
### *(21/4/2005)*

Kyle Williams becomes the new sheriff of Bywater, Louisiana, just as a young man is found dead with plants growing in him. He is not the first – in fact 47 people have gone missing, many to return dead in the same way, and including Williams predecessor, and all since Frederic Schist bought ancient tribal lands from Seminole chieftain Ted Sallis. Joining forces with photographer Mike Ploog, shaman Pete Horn, eco-terrorist Rene LaRoque and schoolteacher Teri Richards, Williams and his deputy try to locate the monster in the swamps seemingly intent on destroying Schist.

**Cast:** Matthew Le Nevez (Sheriff Kyle Williams), Rachael Taylor (Teri Richards), Rawiri Paratene (Pete Horn), Steve Bastoni (Rene LaRoque), Robert Mammone (Mike Ploog), Alex O'Loughlin (Deputy Eric Fraser), Jack Thompson as Schist, *Mark Stevens (Ted Sallis/Man-Thing)*

**Dir:** Brett Leonard; **Writer:** Hans Rodionoff; **Prod:** Avi Arad, Christopher Petzel, Scott Karol, Gimel Everett; **Music:** Roger Mason; **Exec.Prod:** Stan Lee, Kevin Feige, Avi Arad, Rudolf G Wiesmeier, Christopher Mapp; **DOP:** Steve Arnold ACS; **Prod.Des.:** Peter Pound, Tim Ferrier; **Ed:** Martin Connor; **Costumes:** Cappi Ireland; **Lionsgate /Fierce Entertainment/Marvel Enterprises/Screenland Movieworld**; 97; c $6m; $1m

**Notes:** Alongside Mike Ploog, two other characters – Val Mayerik and Steve Gerber – are named for Man-Thing artists/writers. The film was made in Australia, and consequently has a predominantly Australian cast. The film was originally intended to be straight to DVD, but was then decided to be a theatrical release, though after much delay, the film was finally released on the SciFi cable channel in America, and with a limited international theatrical release.

**Comic Notes:** In the comics, Man-Thing is indeed the alter-ego of Ted Sallis, but aside from that the movie and comic characters have virtually no similarity whatsoever. First appearing in *Savage Tales #1 (May, '71)* and created by Stan Lee, Roy Thomas, Gerry Conway and Gray Morrow, Ted Sallis was a biochemist working in the Everglades alongside Bobbi Morse. Sallis – who once worked with Curt Connors – was attempting to recreate the Super Soldier serum. His girlfriend – Ellen Brandt (later retconned to be his wife) – sold him out to the Advanced Ideas Mechanics, and Sallis injected his serum attempt, falling into a swamp, and becoming Man-Thing (though another retcon revealed that magic was also involved). Man-Thing was a lumbering, mute, plant creature who secreted acid, and he defeated AIM and badly scarred his wife, before heading into the swamps. Ted Sallis would be largely lost, though on occasion his personality could surface. Although a definite part of the mainstream Marvel universe, Man-Thing exists in the same sort of space as Howard the Duck and Deadpool; able to break the fourth wall (indeed he encounters two of his writers at various points, one of whom briefly becomes Man-Thing). He currently serves in Phil Coulson's Howling Commandos.

**Ratings:** *IMDB:* 41%; *Rotten Tomatoes:* 17%

**Review:** As an Australian, it wounds me to say that this is largely rubbish. The majority of the Aussie actors struggle to create a believable Louisiana accent, but worse than that, the film genuinely doesn't know what it is. It opens as though it were an eighties horror movie – complete with nudity and gore – but then moves into some sort of strange X-Files like episode. When your title character isn't your main character – Mark Stevens doesn't even get a titles' credit – it's clear the writers don't know what to do with the character. Far too different from its source material, this needed a much more talented production team to make it work.

## FANTASTIC FOUR
### *(8/7/2005)*

Pursuing the possibility that cosmic energy triggered evolution, Reed Richards requests the use of Victor Von Doom's space station to study a cosmic energy cloud. Von Doom agrees – in exchange for profits resulting from any find – and Reed, Von Doom, his assistant and Reed's former girlfriend Sue Storm, her astronaut brother Johnny, and Reed's astronaut friend Ben Grimm, all travel to the station. The cosmic energy has a strange effect on all of them – giving them bizarre super abilities – and as Reed struggles to find a way to restore his friends, they learn to adapt to their new lives and Von Doom, having lost everything thanks to the failure, sets out for revenge on Reed.

**Cast:** *Ioan Grufford (Reed Richards), Jessica Alba (Susan Storm), Chris Evans (Johnny Storm), Michael Chiklis (Ben Grimm), Julian McMahon (Dr Victor Von Doom/Dr Doom), Kerry Washington (Alicia Masters),* Hamish Linklater (Leonard), Laurie Holden (Debbie McIlvane), David Parker (Ernie), Kevin McNulty (Jimmy O'Hoolihan), Maria Menounos (Nurse), *Stan Lee (Willie Lumpkin)*

**Dir:** Tim Story; **Writers:** Michael France, Mark Frost; **Prod:** Avi Arad, Bernd Eichinger, Ralph Winter; **Music:** John Ottman; **Exec.Prod:** Michael Barnathan, Chris Colombus, Kevin Feige, Mark Radcliffe, Stan Lee; **DOP:** Oliver Wood; **Prod.Des.:** Bill Boes; **Ed:** William Hoy;**Costumes:** José I Fernandez; **20th Century Fox/Constantin Film/1492 Pictures/Marvel Enterprises**; 106; c $95m; $333m

**Stan Spotting:** Obviously he's Willie Lumpkin…read on.

**Notes:** Producer Ralph Winter has a cameo during the film as a construction worker towards the end of the film. Chris Columbus was keen to make a Fantastic Four movie, but Bernd Eichinger's Constantin Films held the rights. The 1994 film was made to stop Columbus obtaining the rights, but with funding needed he was brought on board for this one. The DVD release, curiously, is different to the theatrical release, with slightly different edits of certain scenes. In 2007 an extended cut of the film was released, including an extra 20 minutes of footage.

**Comic Notes:** For Reed, Sue, Johnny, Ben, Doom and Alicia's first appearances, see **Fantastic Four (1994)**. There is a nice throwaway reference to Alicia's father; Ben talks about some puppet's which Alicia says were made by her father – in the comics Phillip Masters is the villainous Puppet Master. Victor accompanying the Four on their excursion is actually more similar to *Ultimate Fantastic Four*, which had only just started publishing at the time of production. Though not generally referred to in the film, a post-credits sequence shows Von Doom is indeed from Latveria.

*Willie Lumpkin* is the postman for the Baxter Building, and first appeared in *The Fantastic Four #11 (Feb, '63)*, created by Stan Lee, Dan DeCarlo and Jack Kirby, though in truth there was an earlier version that Lee and DeCarlo had created for a newspaper comic strip in 1960.

**Ratings:** *IMDB:* 57%; *Rotten Tomatoes:* 27%; *Metacritic:* 27

**Review:** This film gets a bad rap, but it's actually not as bad as people suggest. McMahon is woefully miscast, and it's a little difficult to buy Alba as a scientist (think Denise Richards in **The World Is Not Enough**), and the plot is a little thin. However, Grufford, Chiklis and Evans are outstanding, and the chemistry between the four leads is perfect, making the film an easy watch. By the end of the film, even McMahon is bearable. Of the three Fantastic Four attempts, this is easily the one that comes closest to getting it right.

## X-MEN: THE LAST STAND
### *(25/5/2006)*

Worthington Labs reveals to the world it has a cure for mutation, but this infuriates Magneto who reforms his Brotherhood and bolsters its ranks, in anticipation of a war with the humans. Scott goes to Alkali Lake where he finds Jean, who kills him; she is now possessed of an alternate personality called the Phoenix which also kills Xavier, and aligns herself with Magneto. With two obvious threats, Wolverine, Storm and Beast come together to form a new group of X-Men to fight back.

**Cast:** *Hugh Jackman (Logan/Wolverine), Halle Berry (Ororo Munroe/Storm), Ian McKellan (Erik Lehnsherr/Magneto), Patrick Stewart (Professor Charles Xavier), Famke Janssen (Jean Grey), Anna Paquin (Marie/Rogue), Kelsey Grammer (Dr Henry 'Hank' McCoy/Beast), James Marsden (Scott Summers/ Cyclops), Rebecca Romijn (Raven Darkholme/Mystique), Shawn Ashmore (Bobby Drake/Iceman), Aaron Stanford (John Allerdyce/Pyro), Vinnie Jones (Cain Marko/Juggernaut), Ellen Page (Kitty Pryde/Shadowcat), Daniel Cudmore (Peter Rasputin/Colossus), Ben Foster (Warren Worthington III/ Angel), Michael Murphy (Warren Worthington II), Dania Ramirez (Callisto), Shohreh Aghdashloo (Dr Kavita Rao),* Bill Duke (Trask), *Eric Dane (Multiple Man), Cameron Bright (Jimmy/Leech), Kea Wong (Jubilation Lee/Jubilee), Shauna Kain (Theresa Rourke Cassidy/Siryn), Luke Pohl (Flea), Richard Lee (Little Phat), Via Saleaurmua (Phat), Mei Melançon (Psylocke), Omahyra (Arclight), Clayton Watmough (Glob Herman), Ken Leung (Kid Omega), Lance Gibson (Spike), Olivia Williams (Dr Moira MacTaggert)*

**Dir:** Brett Ratner, **Writers:** Simon Kinberg, Zak Penn; **Prod:** Lauren Schuler Donner, Ralph Winter, Avi Arad; **Music:** John Powell; **Exec.Prod:** John Palermo, Kevin Feige, Stan Lee; **DOP:** Dante Spinotti; **Prod.Des.:** Edward Verreaux; **Ed:** Mark Goldblatt, Mark Helfrich, Julia Wong; **Costumes:** Judianna Makovsky, Lisa Tomczeszym; **20th Century Fox/Marvel Entertainment/ Donners' Company/Ingenious Film Partners/Dune Entertainment/Major Studios Partners**; 104; $210m; $460m

**Stan Spotting:** Jean Grey's neighbour when she was very young.

**Notes:** Bryan Singer opted to turn down directing this film in favour of directing **Superman Lives**, a decision he regrets (along with everyone who saw both this movie and **Superman Lives**). Singer's early plans would have seen Jean indeed become the Dark Phoenix, but be recruited by the Hellfire Club. Due to the limited availability of Rebecca Romijn, Anna Paquin and James Marsden (who ironically was busy filming **Superman Lives**), the characters were written out of the movie as soon as possible. Maggie Grace was actually cast as Kitty Pryde until it was decided she was too old for that part.

**Comic Notes:** *Hank McCoy and Warren Worthington III:* make up the other two members of the original X-Men, and as such first appeared in *The X-Men #1 (Sep, '63)*, created by Stan Lee and Jack Kirby. McCoy (who had a cameo in the previous film), or Beast, is quite similar to his movie counterpart, though originally he was a simple human with oversized hands and feet and superhuman strength and agility (which is, of course, much like how he is portrayed in **First Class**). He would leave the X-Men when he turned twenty, and go on to discover a mutagenic serum, which he would self-test, thus turning him blue and furry. Not long after this he would join the Avengers, the Defenders, and then would join his original teammates in forming X-Factor. A later secondary mutation would give him more feline features. Warren Worthington had already embraced his mutation – which is essentially what is seen in the film – to become the Avenging Angel, and was then recruited to the X-Men. After Professor Xavier formed a new X-Men team, Angel and Iceman left to form the Champions, and then with Beast joined the Defenders, before joining X-Factor. During the "Mutant Massacre", Angel's wings were impaled and he entered a state of depression. He was sought out by the mutant Apocalypse, who gave him metal wings, along with changing his physical appearance to become Death. He would later re-join his teammates, now calling himself Archangel, though over time both his blue skin and metal wings have come and gone, replaced for various reasons with the feathered wings and normal skin. He is currently in a relationship with Paige Guthrie.

*Callisto*: first appeared in *Uncanny X-Men #169 (May, '83)* created by Chris Claremont and Paul Smith. She shares most of the movie version's powers, though she cannot sense mutants and their abilities (this was, in fact, the powers for another mutant named Caliban). She is the leader of a group of mutants, like the movie, but in the comics these are the Morlocks. She and Caliban formed the Morlocks as a refuge, but she became somewhat obsessive about them, until she was beaten by Storm in a dual for leadership. She has since tempered her views and the Morlocks are allies to the X-Men now.

*Warren Worthington Jr.* first appeared in *The X-Men #14 (Nov, '65)*, created by Lee, Kirby and Jay Gavin. In the comics he is unaware that his son is a mutant, and indeed when he found out his memory was wiped by Xavier. He was later murdered by his brother Burt, another mutant who called himself Dazzler.

*Kavita Rao*: first appeared in *Astonishing X-Men #1 (Jul, '04 – Vol 3)*, created by Joss Whedon and John Cassaday. Her storyline actually heavily influenced the movie's storyline; she created the Hope Serum to cure mutation and also stop the Legacy Virus which was killing Colossus. When Beast investigated, this led to the X-Men infiltrating Rao's laboratories to destroy the virus. Uniquely, when all mutations were wiped out by the Scarlet Witch, Rao joined the X-Club, a group set up by Beast and Angel with a view to restoring mutations.

*The Multiple Man*: or James Madrox – first appeared in *Giant Size Fantastic Four #4 (Feb, '75)*, created by Len Wein, Claremont and John Buscema. His powers are the same as the movie's, though unlike the comics he is actually an X-Man, working initially with Moira MacTaggart on Muir Island, and then going on to join a later iteration of X-Factor.

*Elizabeth "Betsy" Braddock*: or Psylocke, first appeared in *Captain Britain #8 (Dec, '76)* created by Claremont and Herb Trimpe. Essentially a powerful telekinetic, she is the sister of Brian Braddock – Captain Britain. She once stood in for Brian as Captain Britain, and the resultant battle saw her blinded. She was later abducted by Mojo who gave her bionic eyes, and was rescued by Xavier's New Mutants, whereupon she joined the X-Men. Initially she was an English Rose, in a flowing pink costume, but this became armor when the X-Men apparently died, though in fact went into hiding in Australia. After passing through the Siege Perilous – a gateway to other realities – she was found amnesic in China, and Mojo's assassin Spiral swapped her mind with that of the girl Kwannon. The Chinese version of Psylocke is the one the movie is best based on, though this version wears a very skimpy purple costume. In Kwannon's body, Psylocke is able to generate psychic daggers around her hands. She was killed while fighting Vargas, but was of course resurrected. Incidentally, Kwannon awoke in Betsy's body and became Revanche, and latterly joined the X-Men.

*Arclight*: first appeared in *Uncanny X-Men #210 (Oct, 86)*, created by Claremont. Her real name was Philippa Sontag, and she could indeed generate seismic shockwaves (though she also had superhuman strength). She was a member of the Marauders and assisted with the Mutant Massacre, and has been cloned repeatedly by Mr Sinister, making it unclear what has happened to the original.

*Kid Omega*: first appeared in *New X-Men #134 (Jan, '03)*, created by Grant Morrison and Keron Grant. He is nothing like the movie version – Quentin Quire, as he is more commonly known, has enormous psionic powers and superhuman intellect. Although a member of the X-Men, he was bored by this and became a villain. The movie version of Kid Omega is more like Quill, who first appeared in *New X-Men: Academy X #1 (Jul, '04)*, created by Nunzio de Filippis and Christina Weir. He was part of Cyclops' training team, though he was more interested in learning thievery from Gambit. Later, he was killed by William Stryker.

*Leech*: first appeared in *Uncanny X-Men #179 (Mar, '84)*, created by Claremont and John Romita Jr. He has his movie counterpart's power dampening skills, but looks nothing like him. Instead he is green, tends to have a misshapen skull, and sometimes green lumps. He was taken in by the Morlocks when he was abandoned by his parents, and would become a ward of X-Factor after the Mutant Massacre. Leech became very close friends with Artie Maddicks.

*Spike:* There have been several Spikes in the comics, but the one that the movie version is probably inspired by is the one that debuted in *X-Force #121 (Dec, '01)*, created by Peter Milligan and Mike Allred. Obviously this one can generate bone spikes to throw at people.

*Phat:* debuted four issues earlier in *#117 (Aug, '01)*, also created by Milligan and Allred. William Reilly was a young man who can make his body larger thanks to the ability to expand his fat deposits. Phat was openly gay, which was important to his time with the *X-Force*, though it did create a rift in the team. Both Spike and Phat were killed when the team took on Mister Code's cult.

*Robert "Glob" Herman:* first appeared in *New X-Men #117 (Oct, '01)*, created by Morrison and Ethan Van Sciver, and unlike the movie version, is very much a completely transparent human being. He also possessed superhuman strength, speed and resilience thanks to the bio-paraffin body he has. Despite being led astray by Kid Omega, and briefly joining the Hellfire Academy, Glob is a loyal X-Man.

*Moira MacTaggert:* makes an uncredited cameo in the film. She first appeared in *Uncanny X-Men #96 (Dec, '75)*, created by Claremont and Dave Cockrum. Like this movie version, she is a geneticist, and expert in mutant affairs (very different to the CIA agent in **First Class**). She knew Charles Xavier when he was at Oxford, and the pair were engaged, though they never married. Instead she married Joseph MacTaggert (though his surname has variously been spelled McTaggert, or MacTaggart) who beat and raped her. She left her husband to start a mutant clinic on Muir Island, and would form a close relationship with Sean Cassidy (Banshee). She rescued a girl named Rahne Sinclair and became a foster mother to the girl. It later transpired that she had a second clinic with troubled mutants near Xavier's school, which included Gabriel Summers – the brother to Scott and Alex. The students were actually moulded by Xavier to be a new team of X-Men, set up to save the previous team, but when all were killed, Xavier wiped everyone's memory to forget about them. When the Legacy Virus appeared, Moira worked on a cure, but Mystique altered it to target normal humans and Sabretooth brutally attacked Moira to stop her developing the cure. Moira died in Xavier's arms as she mentally transferred the Legacy cure to him.

*And:* Most of the mutants that appear in the movie are simply cameos, but there are three identical blonde girls that appear at the end of the film. These are supposed to be the Stepford Cuckoos, identical sisters who have the ability to act as a hive mind, and have powerful telepathic abilities and can create a diamond skin. They first appeared in *New X-Men #118 (Nov, '01)*, created by Morrison and Van Sciver. Originally there were five of them – Celeste, Esme, Irma (or Mindee as she prefers), Phoebe and Sophie. Esme left the Cuckoos to join Xorn and his Brotherhood of Mutants, but Xorn killed her when she attacked him for rejecting her advances. She died in Emma Frost's arms, though Emma was most proud of her. Sophie was the most free-thinking of the Cuckoos, and she fought Quentin Quire when he started a riot at the school. However, Esme prompted her to use the drug Kick to increase her powers, and this killed her. Irma has had several failed relationships, but has since dyed and cut her hair to separate herself from her sisters. Celeste is the most compassionate of the Cuckoos, though she has problems with the teenage Jean Grey, and fears losing a connection with her sisters. Phoebe is the most power hungry of the group. In a storyline called "Warsong" it was revealed that the girls are clones of Emma Frost, and that there are actually thousands of them. They are Weapon XIV, designed to collect data on the X-Men, but some of the Phoenix force sought them out to destroy them rather than let the thousands of clones become a deadly force. Celeste embraced the Phoenix force to control it, but it destroyed the clones. Phoebe wanted to continue to keep the Phoenix, but Celeste sealed it in her, Phoebe's and Irma's diamond hearts, stopping them from feeling emotions again.

**Ratings:** *IMDB:* 68%; *Rotten Tomatoes:* 58%; *Metacritic:* 58

**Review:** There is an awful lot wrong with this movie. In some ways I feel sorry for Brett Ratner, who was placed in an unfortunate situation, needing to change the logic of the film to accommodate the changes in circumstances (it should definitely have been Cyclops who killed Jean at the end, not Wolverine), but whilst it's understandable, Ratner makes a lot of dumb choices that destroy the characters set up in the earlier films, and fails to make anything of the new characters. The film tries to be something it's ultimately not, and any punches never truly land. Certainly the weakest of the trilogy, and a waste of some great characters.

## GHOST RIDER
### *(15/1/2007)*

Johnny Blaze makes a pact with the devilish Mephistopheles to hand his own soul over in exchange for his father being cured from cancer. While Barton Blaze does survive his cancer, he later dies in one of his motorcycle stunts. Years later, Blaze rekindles a friendship with Roxanne Simpson, as Mephistopheles' son Blackheart and his fallen angels arrive to take the souls of a village. Mephistopheles, playing the odds, offers Blaze his soul back if he defeats Blackheart. In addition to this, Blaze is given the ability to transform in the Ghost Rider. When Blaze runs into a former Ghost Rider, he learns a few things and decides to not only defeat Blackheart, but also to save every one of the souls.

**Cast:** *Nicholas Cage (Johnny Blaze/Ghost Rider), Eva Mendes (Roxanne Simpson), Wes Bentley (Blackheart), Sam Elliott (Carter Slade),* Donal Logue (Mack), *Peter Fonda (Mephistopheles), Brett Cullen (Barton Blaze),* David Roberts (Captain Dolan), Matt Long (Young Johnny Blaze), Raquel Alessi (Young Roxanne Simpson)

**Dir/ Writer:** Mark Steven Johnson; **Prod:** Avi Arad, Michael de Luca, Gary Foster, Steven Paul; **Music:** Christopher Young; **Exec.Prod:** Norm Golightly, David S Goyer, Lynwood Spinks, E Bennett Walsh, Stan Lee; **DOP:** Russell Boyd; **Prod.Des.:** Kirk M Petruccelli; **Ed:** Richard Francis-Bruce; **Costumes:** Lizzy Gardiner; **Crystal Sky Pictures/Relativity Media/Marvel Entertainment/Michael de Luca Productions/GH One/Vengeance Productions Pty Ltd** ; 110; $110m; $228m

**Notes:** Nicholas Cage is a huge Ghost Rider fan, and pushed for the role in order to get it. To play Johnny Blaze, he needed to have a tattoo covered up – ironically, the tattoo was of Ghost Rider.

**Comic Notes:** *Johnny Blaze*: first appeared in *Marvel Spotlight #5 (Aug, '72)* and he was created by Roy Thomas, Gary Friedrich and Mike Ploog. Ghost Rider's first appearance is slightly more interesting. Magazine Enterprises published a character called Ghost Rider – named for the bluegrass song "Ghost Riders In The Sky", which was created by Ray Krank and Dick Ayers in *Tim Holt #11 ('49)*. When the trademark on the character lapsed, Marvel immediately launched its own Ghost Rider, who debuted in *Ghost Rider #1 (Feb, '67)*, and was created by Gary Friedrich, Roy Thomas and original creator Dick Ayres. This Ghost Rider was Carter Slade. When Friedrich and Thomas created the Johnny Blaze version, Slade's Rider was renamed the Night Rider, and then quickly renamed the Phantom Rider (Night Rider being a term used for a member of the Ku Klux Klan). Johnny Blaze's backstory in the comics is analogous to the movie, though in the comics he is taken in by Roxanne Simpson's family after his father was killed, and he sold his soul to Mephisto in order to save the life of Roxanne's father, Crash.

*Carter Slade*: in the movie is a mysterious figure whose past is never fully explained, while in the comics Slade existed in the wild west, but later possessed his descendent, Hamilton Slade, to bring the Phantom Rider back. His title as Caretaker is a reference to a later Ghost Rider story arc which suggested that there were four Spirits of Vengeance, and the Caretaker existed to watch over the two bloodlines who carried the spirits within them.

*Barton Blaze*: made his only appearance in *Marvel Spotlight #5* as well, and his backstory is broadly in keeping with the movie, having died from a motorcycle accident when Johnny was but a boy.

*Roxanne Simpson*: also first appeared in *Marvel Spotlight #5*, and while her character is similar to the movie version, one of the biggest differences is that Roxanne drove Mephisto away when he tried to take Johnny's soul in return for saving Crash Simpson's life.

*Mephistopheles*: was renamed Mephisto for *Silver Surfer #3 (Dec, '68)*, and was created by Stan Lee and John Buscema. The character was essentially the Mephistopheles envisioned by Faust.

*Blackheart*: first appeared in *Daredevil #270 (Sep, '89)* and was created by Ann Nocenti and John Romita Jr. He is indeed the son of Mephisto (after a fashion), and whilst he menaced Ghost Rider, his principle targets were actually Daredevil and Spider-Man.

**Ratings:** *IMDB:* 52%; *Rotten Tomatoes:* 26%; *Metacritic:* 35

**Review:** An odd film, this one. There's an awful lot to like about it, and Ghost Rider is given some wonderful menace, but at times he feels a little disjointed, and his powers seem a little choosy about when and what they should do. Eva Mendes is extremely likable, and Peter Fonda is classy all the way, but Wes Bentley lacks any real threat, and Nicholas Cage…well, for a guy who was such a Ghost Rider fan, it's bizarre that he never really plays the character in any recognisable way. Certainly not on par with the majority of the Marvel output.

## SPIDER-MAN 3
### *(16/4/2007)*

As Peter sets out to finally marry Mary Jane, several forces begin working against him. Flint Marko falls into a particle accelerator, turning his body into sand. Harry Osborn takes on the mantle of the Goblin to exact revenge against Spider-Man, who he now knows is Peter. And an alien symbiote crash lands on the planet, attaching itself to Peter and enhancing his negative emotions.

**Cast:** *Tobey Maguire (Spider-Man/Peter Parker), Kirsten Dunst (Mary Jane Watson), James Franco (New Goblin/Harry Osborn), Thomas Haden Church (Sandman/Flint Marko), Topher Grace (Venom/Eddie Brock), Bryce Dallas Howard (Gwen Stacy), Rosemary Harris (May Parker), J K Simmons (J Jonah Jameson), James Cromwell (Captain George Stacy),* Theresa Russell (Emma

Marko), *Dylan Baker (Dr Curt Connors), Bill Nunn (Joseph "Robbie" Robertson),* Bruce Campbell (Maître d'), *Elizabeth Banks (Betty Brant),* Ted Raimi (Hoffman), Perla Haney-Jardine (Penny Marko), *Willem Dafoe (Green Goblin/Norman Osborn), Cliff Robertson (Ben Parker)*

**Dir:** Sam Raimi; **Writers:** Sam Raimi, Ivan Raimi, Alvin Sargent; **Prod:** Laura Ziskin, Avi Arad, Grant Curtis; **Music:** Christopher Young; **Exec.Prod:** Stan Lee, Kevin Feige, Joseph M Caracciolo; **DOP:** Bill Pope ASC; **Prod.Des.:** Neil Spisak, J Michael Riva; **Ed:** Bob Murawski; **Costumes:** James Acheson, Katina Le Kerr; **Columbia Pictures/Marvel Entertainment**; 139; c $300m; $895m

**Notes:** Though Christopher Young takes on the music for this film, Danny Elfman is still credited for "original music themes". Both Young and Grant Curtis have cameo appearances in the film. This is one of the most expensive films made in the US (indeed, some have argued that taking into account the development of new technology for the film, it is *the* most expensive). Sadly, this was Cliff Robertson's last film.

**Stan Spotting:** The "man in Times Square" who says to Peter that one man can make a difference. 'Nuff said!

**Comic Notes:** *Gwen Stacy*: first appeared in *The Amazing Spider-Man #31 (Dec, '65)*, another Stan Lee/Steve Ditko creation. She had a crush on Peter after meeting him when they both attended Empire State University, but would date both Flash Thompson and Harry Osborn before Peter ended his relationship with Mary Jane and started dating her. In *The Amazing Spider-Man #121 (Jun, '73)* the Green Goblin threw her off a tower, and Peter used his webbing to grab her, though the whiplash of the grab killed her. In this movie, obviously, she is nothing more than a second love interest for Peter, but she has more character in **The Amazing Spider-Man**, and her death is closely followed in **The Amazing Spider-Man 2**. Gwen's death has, surprisingly, remained locked in, but several clones of her have turned up, and the, frankly bizarre revelation that she had twins to Norman Osborn was also introduced. Recently, a parallel universe where Gwen became Spider-Woman was explored by the comics, and was so popular the miniseries became an ongoing one; *Spider-Gwen*.

    *George*: Gwen's father, debuted in *#56 (Jan, '68)*, this time created by Lee, John Romita Sr and Don Heck. Thirty-four issues later in *#90 (Nov, '70)* he was also killed off. In the comics Stacy is a retired cop, though he keeps up with NYPD, and took a particular interest in Spider-Man, ultimately deducing who he was. Again he isn't particularly explored in this movie, but his death in **The Amazing Spider-Man** is rather different to the comics, in which he saves a child from being crushed by debris. Like the movie, however, he does beg Peter to look after Gwen, given his lifestyle.

*William Baker*: who would later adopt the name Flint Marko, and then go onto become the Sandman – was one of the earliest Spider-Man villains, turning up in *The Amazing Spider-Man #4 (Sep, '63)*, and if you guessed he was created by Lee and Ditko, you can claim a gold star. His background is broadly the same as the movie version; Marko strayed into a nuclear testing site, and his body bonded with irradiated sand. He was already a villain, so he simply used his new abilities to continue this line of work. A large number of different storylines have featured the character, showing his powers altering over time, but none of them have featured a wife or child.

The next one is a little complicated…

*Eddie Brock*: first appeared in *Web of Spider-Man #18 (Sep, '86)*, created by David Michelinie and Marc Silvestri. Spider-Man gained a new costume in *The Amazing Spider-Man #252 (May, '84)*, which was written by Tom DeFalco and Roger Stern, with art by Ron Frenz. However, the costume was actually created by fan Randy Schueller, and was designed, from Schueller's idea, by Mike Zeck. Aside from the black costume idea turning up in this movie, what has this to do with Eddie Brock? In *Secret Wars #8 (Dec, '84)* it was revealed that during this time, Peter Parker's costume was damaged and he was given a new, black costume, that was actually an alien symbiote. Like the movie, when Peter realised that symbiote was alive and susceptible to sound, he used a cathedral bell to get rid of the costume. Eddie Brock, on the other hand, was a journalist who disgraced himself by exposing a killer who was actually nothing more than a compulsive liar. Blaming Spider-Man, Brock goes to a church, and the symbiote attaches to him, creating the villain Venom – debuting in *The Amazing Spider-Man #299 (Apr, '88)*, which was written by Michelinie, with art by Todd McFarlane. As such, the actual creator of Venom is difficult to accurately credit. Michelinie once claimed sole credit when a magazine suggested it was he and McFarlane who created the character, and Peter David, who wrote *The Spectacular Spider-Man* storyline in which Eddie Brock's career dive occurred, completely backed this up. Writer Erik Larsen suggested that Michelinie simply stole an idea that had already been created, and further indicated that McFarlane's work made the character truly what it was. As such, it seems Venom is a genuinely group created character. As Venom, Brock would fight Spider-Man, but then fight alongside him against other symbiotes such as Carnage, become something of an anti-hero, before returning to a life of crime. The symbiote would bond with another, and then attempt to return to Brock who had, by now, contracted cancer, and the bonding would result in the Anti-Venom. A large number of other symbiotes exist, and Brock was forced to bond to one named Toxin. As Toxin, Brock would become a vigilante, rather than a villain.

*And:* the Carjacker's name is given as Dennis Carradine in this film. In the comics he does have a daughter named Jessica Carradine, so while a name is never given to the carjacker, there is the implication his surname is Carradine.

**Ratings:** *IMDB:* 62%; *Rotten Tomatoes:* 63%; *Metacritic:* 59

**Review:** Outside of MCU Marvel films, it could be argued that the third in the trilogy is the slip up for most of the series – X-Men, Blade and, inevitably, Spider-Man. The film really doesn't work, and there's plenty of people who point the blame at the studio, but in fairness, they aren't responsible for Peter Parker's emo dance sequence, which is easily the nadir of the Spider-Man series. Venom is badly undeveloped, and the New Goblin doesn't make enough of an impact as a villain to care about the fact he turns good; indeed Harry's amnesia seems like an easy way to get past an annoying plot problem. Everything that was good about the previous film is bad in this one, and Raimi and his team go out on a low.

## Spider-Man 4?

Sam Raimi had intended to do a fourth Spider-Man film, but his experiences with the making of the third led him to part with Colombia and give up on the series. However, since that time, we've learnt of what we could have had with the fourth film...

In fact, the fourth film had made quite some headway. Tobey Maguire was the only cast member officially signed on for the film, though Raimi expressed his desire for both Kirsten Dunst and Bruce Campbell to return as well, though Dunst's return seemed very uncertain. James Vanderbilt wrote the original screenplay, but David Lindsay-Abaire and Gary Ross had also performed rewrites, so a significant script was developed. The film was given a release date of May, 2011.

A number of ideas for the fourth film ultimately would appear in **The Amazing Spider-Man** films, but surprisingly the Lizard wasn't one of them. Despite Dylan Baker's recurring appearance as Curt Connors, it seemed that Raimi was keen on the idea of using Electro, Vulture and possibly the Sinister Six. John Malkovich was rumoured to be in line to play Vulture, and Anne Hathaway was potentially going to be Felicia Hardy – the Black Cat (oh, the irony). One of the concept artists for the film had also come up with artwork of Mysterio.

Ultimately, of course, none of this would come to pass. Sony would reboot the series, Raimi would walk away, and a script has yet to surface to show us what was actually planned. Whether it would have gone back to the quality of the second film or continued the downward trend of the third is anyone's guess.

## FANTASTIC FOUR: RISE OF THE SILVER SURFER
### (12/6/2007)

Despite an imminent wedding, Reed agrees to help the Army investigate the arrival of a silver thing that creates craters around the world. During the wedding, the Silver Surfer zooms past, but when Johnny follows it, it creates instability with his – and the others' – powers. In Latveria, the Surfer releases and heals Doom, who then goes to the American Army with a deal, and soon the Fantastic Four are forced to work with Doom to capture the Surfer. However, once captured, the Surfer reveals that he is the herald for a being of great power – Galactus – and Galactus' intends to feed on the planet itself.

**Cast:** *Ioan Grufford (Reed Richards/Mr Fantastic), Jessica Alba (Susan Storm/Invisible Woman), Chris Evans (Johnny Storm/Human Torch), Michael Chiklis (Ben Grimm/The Thing), Julian McMahon (Dr Victor Von Doom/Dr Doom), Kerry Washington (Alicia Masters),* Andre Braugher (General Hagar) *with Laurence Fishburne as the Voice of the Silver Surfer, Doug Jones (The Silver Surfer), Beau Garrett (Captain Frankie Raye),* Brian Posehn (Wedding Minister), Zach Grenier (Mr Sherman), Kenneth Walsh (Dr Jeff Wagner), Vanessa Minnillo (Julie Angel), Kevin McNulty (Jimmy O'Hoolihan)

**Dir:** Tim Story; **Writers:** John Turman, Mark Frost, Don Payne; **Prod:** Bernd Eichinger, Avi Arad, Ralph Winter; **Music:** John Ottman; **Exec.Prod:** Stan Lee, Kevin Feige, Chris Colombus, Mark Radcliffe, Michael Barnathan; **DOP:** Larry Blanford; **Prod.Des.:** Kirk M Petruccelli; **Ed:** William Hoy ACE, Peter S Elliot; **Costumes:** Mary Vogt; **20th Century Fox/Constantin Film/1492 Pictures/ Marvel Entertainment**; 92; $125m; $301m

**Stan Spotting:** Himself being rejected from Reed and Sue's wedding.

**Notes:** General Hagar was originally going to be Nick Fury. Initially it was planned that the Silver Surfer would be mute and Galactus – more akin to the comic version – could speak, and would be voiced by Fishburne. When the decision was taken to become more like the Ultimate Marvel comics, Fishburne swapped roles. Hype suggested that Fox were keen to pursue a

Silver Surfer spin-off film, but this was ultimately decided against.

**Comic Notes:** *Norin Radd*: the Silver Surfer – first appeared in *Fantastic Four #48 (Mar, '66)* and was created by Jack Kirby. His story is broadly similar to this movie – he was the herald for Galactus but turned against the creature after meeting the Fantastic Four and Alicia Masters. In the comics, however, they simply drive Galactus away.

*Galactus*: first appeared in the same issue, though he was created by Kirby and Stan Lee (Lee came up with the storyline, but when Kirby sketched out the issue, he created the Surfer). There is a significant difference between the comic version and the movie version – Galactus was formerly Galan, who existed before the Big Bang. During his Big Crunch, he was transformed into Galactus and released with the Big Bang. Huge and hungry, he devours planet simply to survive. He has had a number of heralds over the years, with the second being Radd. The eighth was Nova. Nova was formerly Frankie Raye.

*Frankie Raye*: first appeared in *Fantastic Four #164 (Nov, '75)* and was created by Roy Thomas and George Pérez. She was a United Nations interpreter, though she did fall for, and date, Johnny Storm. In *Fantastic Four #244 (Jul, '82)*, writer John Byrne turned her into Nova and sadly she was killed off a decade later.

*And: Ultimate Extinction (Mar – Jul, '06)* was a comic series written by Warren Ellis, with art by Brandon Peterson, and was a massive crossover event for the Ultimate Marvel universe which dealt with the arrival of Gah Lak Tus. Gah Lak Tus is a large hive of robots with a group mind that swarm and destroy planets. Their arrival is preceded by a number of Silver Surfers who deposit giant beacons in craters around the world. It's not too much of a leap to realise that the version of Galactus in this film is based more on the Ultimate version (indeed some of the dialogue appears verbatim). This was, however, a relatively late change as concept art shows Galactus in his traditional form.

**Ratings:** *IMDB:* 56%; *Rotten Tomatoes:* 37%; *Metacritic:* 45

**Review:** Again a film that has a bad reputation, but again it's harsher than the reality. The cast are once again on top form, and the interaction between them makes it more than worthwhile. New cast additions – particularly Fishburne, Jones, Braugher and Garrett are brilliant. Even McMahon seems better in this film. Perhaps the biggest fault with the film is that the Four are sometimes sidelined for the Surfer, and the even balance of the four in the previous film is altered to give Chris Evans more screen time. However, the film retains a sense of fun, and like its predecessor, is a great family film. For those who wished the series would be more serious...be careful what you wish for.

## PUNISHER: WAR ZONE
### *(5/12/2008)*

The Punisher brings down Billy "The Beaut" Russotti and his brother Loony Bin Jim, but in the process Billy falls into a glass crusher, disfiguring him. During the operation, however, an undercover agent – Donatelli – is killed, much to Castle's horror. While Castle tries to make amends with Donatelli's family, Donatelli's partner joins the NYPD Punisher Taskforce to bring Castle to justice – little knowing that Martin Soap, the only detective on the taskforce, is in league with Castle. As Billy – now calling himself Jigsaw thanks to the repair work on his face – frees his brother and sets out to find Donatelli's family, Castle's armourer – Microchip – convinces Castle he needs to remain as the Punisher in order to protect them.

**Cast:** *Ray Stevenson (Frank Castle), Dominic West (Billy/Jigsaw)*, Julie Benz (Angela), Colin Salmon (Paul Budiansky), Doug Hutchison (Loony Bin Jim), *Dash Mihok (Martin Soap), Wayne Knight (Micro), Mark Camacho (Pittsy), Keram Malciki-Sánchez (Ink), TJ Storm (Maginty), David Vadim (Cristu Bulat), Aubert Pallascio (Tiberiu Bulat), Carlos Gonzalez-Vio (Carlos Cruz)*

**Dir:** Lexi Alexander; **Writers:** Matt Holloway, Art Marcum, Nick Santora; **Prod:** Gale Anne Hurd; **Music:** Michael Wandmacher; **Exec.Prod:** Oliver Hengst, Ernst-August Schneider, Ari Arad, Ogden Gavanski, Michael Paseornek, John Sacchi; **DOP:** Steve Gainer ASC; **Prod.Des.:** Andrew Neskoromny; **Ed:** William Yeh; **Costumes:** Odette Gadoury; **Lionsgate/Marvel Studios/Valhalla Motion Pictures/MHF Zweite Academy Film**; 103; $35m; $10m

**Notes:** Kurt Sutter wrote the first draft of the film, but was so disgusted with the direction the studio wanted to go in, he refused to be credited on the final film. Similarly, Thomas Jane also refused to return because of the film's new direction. The film is similar to **The Incredible Hulk**, in that it is – to use Gale Anne Hurd's terminology – a requel; both something of a sequel and a reboot. The timeline for the film works (about five years after the previous one), though the origin story has been changed – again not actually important, simply, in this case, a couple of lines changing what happened. This is the first film that uses the Marvel Knights logo to open the film. This is, consequently a Marvel Studios film, but it's not part of the MCU, and was distributed by Lionsgate domestically, and Columbia Pictures internationally.

**Comic Notes:** *Billy Russo*: or Jigsaw, first appeared in *The Amazing Spider-Man #161 (Oct, '76)* created by Len Wein and Ross Andru. There's a lot of similarities between the comic and movie versions, including the way Russo became Jigsaw and his nickname "the Beaut", though in the comics, Russo had a wife and son – Henry – rather than a brother. Despite a vendetta against the Punisher, Russo has actually confronted a number of superheroes, and worked with a number of supervillains. Ironically, Henry Russo would become one of the Punisher's allies. Most recently he joined the Kingpin's attempt to rebuild Fisk's criminal empire.

*Linus Lieberman*: credited as Micro, for his comic alias, first appeared in *The Punisher #4 (Nov, '87)*, created by Mike Baron and Klaus Janson. He is much like his film counterpart, though sadly was killed after parting ways with Castle (not, incidentally, by the Punisher). Recently he was resurrected by the Hood, who asked him to kill Castle, which Micro agreed to, which led to him being captured by Jigsaw who handed him over to the Punisher. Fortunately Jigsaw's son saved him before Castle could kill him.

*Martin Soap*: similar to many of the characters in the previous movie, first appeared in Garth Ennis and Steve Dillon's run, *The Punisher #2 (May, '00)* [NB. There have been different volumes of *Punisher*, over the years, with the 1987 run being Volume 2, and the 2000 run being Volume 4]. Soap did indeed head the Punisher Taskforce, but he was fairly corrupt, and ultimately got to the position of Police Commissioner, before karma got him. He ended up working in pornography.

*Carlos Cruz*: was created by Chuck Dixon and Rod Whigham in *The Punisher Vol 2 #97 (Dec, '94)* and like his movie counterpart worked with Micro. However, this was at a point where Micro and the Punisher had fallen out, and Cruz actually took on the mantle of Punisher. Sadly, Micro was killed by a Bolivian named Stone Cold, and Cruz fought Castle, believing him to be the actual killer. When Cruz paused in his fight, Stone Cold killed him.

*Pittsy and Ink*: are Carmine Gazzera and a man named because his first kill was with a pen through the eye. They first appeared in *The Punisher #3 (Apr, '04)* [NB 2004/5 was the Volume 6 run]. Created by Garth Ennis and Lewis Larosa, they were both assassins, though in this case they worked for Nicky Cavella.

*Maginty*: first appeared four issues later - #7 (Aug, '04) – created by Ennis and Leandro Fernandez. Again, he was broadly similar to his movie counterpart; a gang boss.

*Cristu & Tiberiu Bulat*: Cristu first appeared in *The Punisher #25 (Nov, '05)*, created by Ennis and Fernandez; a Romanian brothel owner who suggested to his father they should go into human trafficking. Tiberiu appeared the following issue (#26 (Dec, '05) – Ennis and Fernandez), where he set about trying to eliminate his competition and the Punisher. Needless to say it didn't end well for the Bulats.

And: *The Punisher: War Zone*, was not a storyline, as such, but rather the title of an ongoing comic series during the early nineties, and a follow-up limited series in 2009.

**Ratings:** *IMDB:* 60%; *Rotten Tomatoes:* 27%; *Metacritic:* 30

**Review:** It genuinely astonishes me that this movie was not as well received as the 2004 version. Quite aside from the fact Ray Stevenson is perfect as the Punisher, the film has a much better villain in the form of Jigsaw, and the violence and brutality are stronger and more what you would expect from a Punisher film. This is easily the best of the three Punisher films, and a good film in its own right.

## X-MEN ORIGINS: WOLVERINE
### *(9/4/2009)*

James Howlett watches his father killed by the groundskeeper, which activates his mutation – bone claws – and he kills in revenge; though it transpires the groundskeeper is his biological father. Fleeing with his half-brother Victor – also a mutant with similar abilities – the pair fight war after war, their healing factors keeping them alive. During Vietnam, Major William Stryker makes them part of Team X, but James can't deal with the casual regard for life the team has. Six years later, James – now calling himself Logan – lives with his girlfriend Kayla Silverfox. But when Kayla is found dead, Stryker reveals that Victor has gone rogue, and after going through Stryker's Weapon X program, Logan sets out to find and bring in Victor, little knowing that Stryker has sent out another mutant to kill them both – Deadpool.

**Cast:** *Hugh Jackman (Logan/Wolverine), Liev Schreiber (Victor Creed), Danny Huston (Stryker), will.i.am (John Wraith), Lynn Collins (Kayla Silverfox), Kevin Durand (Fred Dukes), Dominic Monaghan (Chris Bradley/Bolt), Taylor Kitsch (Remy LaBeau), Danniel Henney (Agent Zero), Ryan Reynolds (Wade Wilson), Tim Pocock (Scott Summers), Julia Blake (Heather Hudson), Max Cullen (Travis Hudson), Troye Sivian (James), Michael James Olsen (Young Victor), Peter O'Brien (John Howlett), Aaron Jeffrey (Thomas Logan), Alice Parkinson (Elizabeth Howlett)*

**Dir:** Gavin Hood, **Writers:** David Benioff, Skip Woods; **Prod:** Hugh Jackman,

John Palermo, Lauren Schuler Donner, Ralph Winter; **Music:** Harry Gregson-Williams; **Exec.Prod:** Richard Donner, Stan Lee; **DOP:** Donald M McAlpine; **Prod.Des.:** Barry Robison; **Ed:** Nicholas De Toth, Megan Gill; **Costumes:** Shareen Beringer, Louise Mingenbach; **20th Century Fox/Marvel Entertainment/Dune Entertainment/Donners' Company/Seed/Ingenious Film Partners/Big Screen Productions**; 107; $150m; $373m

**Notes:** April Elleston-Enahoro was cast as a very young Ororo Munroe, but her scenes were ultimately deleted. Patrick Stewart, on the other hand, does indeed cameo uncredited as Professor Xavier towards the end of the film. Deadpool is actually portrayed by Scott Adkins, uncredited. Tahyna Tozzi plays the part of Kayla's sister, Emma, who has the ability to make her body go diamond hard. Until the release of **First Class**, it was assumed this was Emma Frost, but that clearly is not the case, though interestingly Asher Keddie's character is named Dr Carol Frost. Jason Stryker is also briefly seen in the film. Real life poker player Daniel Negreanu is one of LaBeau's opponents during his poker game. Although uncredited, both Avi Arad and Bryan Singer were executive producers on the film. Uniquely, the film has two different post-credit sequences, which were randomly assigned to the cinema you went to. The film was actually shot in Australia to better suit Jackman.

**Comic Notes:** Clearly a large number of characters recur (even briefly) from previous films, so a lot of the comic notes from earlier X-films also apply here.

    *John Wraith*: known also in the comics as Kestrel – first appeared in *Wolverine #60 (Sep, '92 – Vol 2)* and was created by Larry Hama. There's a lot of similarity between his appearance in the movie and the comics, and he was indeed a member of Team X, with teleportation abilities. The comic version's paranoia was also transferred to the movie. One major difference was that Wraith was part of Weapon X, and was eventually hunted down by Sabretooth and almost killed. Later, now a preacher, Wraith again encountered Wolverine though he was possessed by a demon. Hellverine removed his powers and killed his parishioners.

    *Kayla Silverfox*: is, in the comics, Silver Fox, who occasionally uses the name Kayla. She first appeared in *Wolverine #10 (Aug, '89 – Vol 2)*, and was created by Chris Claremont and John Buscema. A Native American, she was indeed Wolverine's lover when he lived in Canada, and was also apparently murdered by Sabretooth, but in fact was not. The comic version was recruited to Team X (as she had the ability to control people by touch), though she betrayed them and Wolverine for Hydra. Later she would come up against Clan Yashida and cause the poisoning of Mariko Yashida. When John Wraith reunited Team X as the age suppressants they had started to fail, Silver

Fox was among them, but was killed when they confronted Sabretooth. Logan buried her in their Canadian home.

*Fred Dukes*: is better known in the comics as Blob, and first appeared in *The X-Men #3 (Jan, '64)*, created by Stan Lee and Jack Kirby. He had a personal gravity field and impenetrable skin, along with superhuman strength and endurance. He was invited to join the X-Men when Xavier discovered he was a mutant, but instead joined Magneto's Brotherhood. He has remained a member of various Brotherhood's, including Mystique's, and was also part of her Freedom Force. He was depowered, though is physical form remained obese, causing depression which led to attempted suicide, but Mystique has since managed to return his powers to him.

*Chris Bradley*: aka both Bolt and Maverick – first appeared in *X-Men Unlimited #8 (Oct, '95)* and was created by Howard Mackie, Tom Grummett and Dan Lawlis. Bradley was recruited to the X-Men by Jean Grey and Gambit when Bradley's electrokinesis manifested and caused him headaches. He was, however, infected with the Legacy Virus, which left him very little time to live. He found a mentor in Maverick, another Legacy sufferer, and, when Maverick apparently died, he took on the identity of Maverick and joined Cable's underground. He was killed, and died in the arms of Agent Zero.

*David North* or Christoph Nord, first appeared in *The X-Men #5 (Feb, '92 – Vol 2)*, and was created by Jim Lee. He had the ability to absorb kinetic energy, as well as a natural healing factor and, curiously, could generate an anti-healing factor corrosive. A German who was part of the Black Ops group Cell Six, he would marry an Italian nurse, little realising she was a spy systematically killing the Cell Six operatives. Nord was forced to kill her in self-defence, and when she revealed he also had killed his unborn child, Nord developed a deep distrust of women. He became a more efficient spy, and the CIA recruited him to Team X, where he changed his name to David North. He would later lose his powers but continue in Team X, adopting the title Maverick. He became a mercenary and then contracted the Legacy Virus, and would later meet the aforementioned Chris Bradley who he became a mentor to. When the Weapon X project was revitalised, North agreed to take part as it would cure him of the Legacy Virus, and he became Agent Zero. Ironically on one mission he encountered a new Maverick who attacked him. Zero killed him, but the pair revealed their true identities and Zero realised that Bradley had attacked to avenge him against the Weapon X project. Zero then took on Bradley's mission to end Weapon X, though the operation had shut down. He was later depowered by Scarlet Witch.

*John and Elizabeth Howlett and Thomas Logan* all made their first appearances in *Origin #1 (Nov, '01)*, created by Paul Jenkins, Bill Jemas, Joe Quesada and Andy Kubert. As the movie, Elizabeth had an affair with Thomas

Logan, and James Howlett was possibly the result of that, though it's never made explicit. Thomas did have a son as well – though it wasn't Victor Creed, and John and Elizabeth had an older son who seemed to have similar abilities to James. John was forced to kill the younger John when he attacked his mother, though the exact circumstances of the situation are unclear. Equally Thomas did kill John in a drunken rage. As such, the storyline for the movie can be seen to come from a combination of these various sources.

*Remy LaBeau*: or Gambit first appeared in *The Uncanny X-Men Annual #14 (Jul, '90)* created by Claremont and Jim Lee (though it was just a cameo – he showed up properly the following month in *The Uncanny X-Men #266)*. He has the ability to convert kinetic energy and transfer it to objects, though in his earlier appearances he could foretell the future with Tarot cards, and control heat. As a teenager he joined the Thieves Guild and was employed by Nathaniel Essex (or Mr Sinister) to retrieve his diaries from the Weapon X project. The fallout from this saw him exiled from New Orleans and the Guild, and he turned to Sinister for protection, which led him to become involved in the Mutant Massacre, something he kept hidden from the X-Men for a considerable period. Gambit left Sinister's employ after the Massacre, and found a depowered Storm who he helped. This saw him join the X-Men, where he formed a relationship with Rogue. When she discovered his involvement in the Mutant Massacre he was kicked out of the team, though she and the team regretted this and sought him out to get him to re-join, which he did. Gambit has subsequently been involved with Sinister and Apocalypse, but eventually returns to the X-fold and Rogue; currently serving as a member of X-Factor.

*Wade Wilson*: or Deadpool, is...a unique character in the Marvel pantheon. First appearing in *New Mutants #98 (Feb, '91)*, and created by Fabian Nicieza and Rob Liefield, he has a healing factor and is an expert martial artist – just like this movie version. The Merc with the Mouth was first seen attempting to kill Cable and the New Mutants, but he was a surprise hit, and soon got his own series. Writer Joe Kelly changed the character slightly, making him less a villain, but more a parody of anti-hero comics of the time. Subsequent writer Christopher Priest even went so far as to make Deadpool appear a little stupid at times. What's interesting about the character is that he is self-aware, and revisionist, so that his back story changes depending on what the writer's want – something Deadpool has pointed out to the audience. It is unclear, for instance, if Deadpool is actually Wade Wilson, or someone who appropriated that identity, and while some stories tell of him being the deranged son of a war hero, others tell of him joining the Weapon X program (as Weapon XI) after being part of Special Forces. No matter what, however, he has always been depicted as hideously deformed underneath his mask. Despite his behaviour and irreverence, Deadpool does seem to have a personal morality – he has never cashed the cheques he received for his assassinations, and he could not reconcile the murder of Apocalypse as a child.

And: There is a Heather Hudson in the comics – much younger she was married to James MacDonald Hudson. She is completely different to her movie counterpart, in point of fact a superhero herself going by the name Vindicator or Guardian (her husband, doing exactly the same thing). Both get their powers of flight, force field generation and the ability to manipulate geothermal energy from a battle suit. They formed Department H in Canada, which involved the creation of superhero team called Alpha Flight. Heather took over leadership of Alpha Flight when James was apparently killed. She first appeared in *The Uncanny X-Men #139 (Nov, '80)* and was created by Claremont and John Byrne.

**Ratings:** *IMDB:* 67%; *Rotten Tomatoes:* 38%; *Metacritic:* 40

**Review: Origins** is complicated by the fact that Hugh Jackman is so perfect as Wolverine, and has such a charismatic and powerful on-screen presence, that it's hard to dislike any film he's in, regardless of the quality of the film. Unfortunately, this movie doesn't do itself any favours with a fairly murky story and some dreadful performances; capped by the most criminal use of Deadpool ever. There are too many elements in the film, and too many 'moments' meaning that most don't really get developed. It's great to watch the first time, but if you go back a second time, it starts to lose its shine.

## Best of the Best/Worst of the Worst

**X-Men: Origins** is largely reviled as one of the worst films superhero films committed to celluloid, but obviously the fun question is, genuinely what are the best and what are the worst? Well, we have three scores for each film, and this author's personal opinion, so... **Captain America (1990)** and **Man-Thing** are lucky to get a pass, as Metacritic hasn't scored them.

The Best of the Best would appear to be:

| | IMDb | Rotten Tomatoes | Metacritic | Author |
|---|---|---|---|---|
| 1...the best! | Deadpool | Logan | Spider-Man 2 | Spider-Man 2 |
| 2 | Logan | Spider-Man 2 | Logan | X2 |
| 3 | X-Men: Days of Future Past | X-Men: Days of Future Past | X-Men: Days of Future Past | Deadpool |
| 4 | X-Men: First Class | Spider-Man | Spider-Man | Spider-Man |
| 5 | X2 | X-Men: First Class | X2 | X-Men: Days of Future Past |

So there's a fair consistency to what should be in the top 5, with **Spider-Man 2** looking like it might be the best non-MCU Marvel movie made. As for the worst....counting down...

| | IMDb | Rotten Tomatoes | Metacritic | Author |
|---|---|---|---|---|
| 5 | Ghost Rider | Blade Trinity | Ghost Rider: Spirit of Vengeance | Spider-Man 3 |
| 4 | Elektra | Ghost Rider: Spirit of Vengeance | Punisher: War Zone | X-Men: Apocalypse |
| 3 | Howard the Duck | Howard the Duck | Howard the Duck | Elektra |
| 2 | Ghost Rider: Spirit of Vengeance | Elektra | Fantastic Four (2005) Fantastic Four (2015) | Ghost Rider: Spirit of Vengeance |
| 1...the worst... | Fantastic Four (2015) | Fantastic Four (2015) | | Fantastic Four (2015) |

Again, a fair degree of consistency, with **Fantastic Four (2015)** seemingly taking out the prize for the worst Marvel film ever made. The cast would appear to agree with that assessment...

## X-MEN: FIRST CLASS
### *(25/5/2011)*

In 1944, a young Erik Lehnsherr watches Klaus Schmidt kill his mother to manifest his powers. They do, and Lehnsherr escapes, but vows revenge on Schmidt. At much the same time, Charles Xavier meets Raven Darkholme, both discovering they have mutant abilities. In 1962, Lehnsherr continues his hunt, and Xavier finishes his studies on mutation. CIA agent Moira MacTaggert uncovers the Hellfire Club, led by Schmidt – now calling himself Sebastian Shaw – and discovers a plan by the Club to start World War III. MacTaggert and her CIA superior seek out Xavier's help, and when they find Shaw they meet Lehnsherr, who agrees to help them. Assembling a team of mutants to take on Shaw's own team, Xavier and Lehnsherr find themselves at philosophical odds, and Raven questions Xavier's methods.

**Cast:** *James McAvoy (Charles Xavier), Michael Fassbender (Erik Lehnsherr), Kevin Bacon (Sebastian Shaw), Rose Byrne (Moira MacTaggert), Jennifer Lawrence (Raven/Mystique), January Jones (Emma Frost), Nicholas Hoult (Hank/Beast), Oliver Platt (Man In Black Suit), Jason Fleming (Azazel), Lucas Till (Alex Summers/Havok), Edi Gathegi (Darwin/Edwin Muñoz), Caleb Landry Jones (Cassidy/Banshee), Zoë Kravitz (Angel Salvadore), Álex González (Janos Quested/Riptide), Don Creech (William Stryker), Matt Caven (CIA Director McCone), Michael Ironside (US Navy Captain), James Remar (US*

General), Ray Wise (US Secretary Of State)

**Dir** Matthew Vaughn; **Writers:** Ashley Edward Miller & Zack Stentz, Jane Goldman & Matthew Vaughn, Sheldon Turner, Bryan Singer; **Prod:** Lauren Schuler Donner, Bryan Singer, Simon Kinberg, Gregory Goodman; **Music:** Henry Jackman; **Exec.Prod:** Stan Lee, Tarquin Pack, Josh McLaglen; **DOP:** Josh Mathieson; **Prod.Des.:** Chris Seagers; **Ed:** Eddie Hamilton, Lee Smith; **Costumes:** Sammy Sheldon; **20th Century Fox/Dune Entertainment/Bad Hat Harry Productions/Donners' Company/Marvel Entertainment/ Ingenious Media/Big Screen Productions**; 131; c $150m; $353m

**Notes:** Both Bryan Singer and Zak Penn were in line to direct this film, though the former turned it down (but did stay on to produce), while the latter dropped out when ideas changed. For a long time after **X-Men: The Last Stand**, alongside **X-Men Origins: Wolverine**, there were plans for a similar movie about Magneto. Writer Sheldon Turner developed this idea a considerable way, envisioning a movie that was **X-Men** crossed with **The Pianist**. However, as ideas progressed, it was thought to change the direction, and Bryan Singer started developing a different type of prequel thanks to an idea from Simon Kinberg (ironically this new idea was similar to the one that Zak Penn had started with). Hugh Jackman cameos as Wolverine, making him the only actor to appear in every **X-Men** film to date, and Rebecca Romijn also cameos as an older version of Mystique. Typically a number of actresses were approached for the part of Emma Frost, with Tahyna Tozzi considered to reprise the role from **X-Men Origins: Wolverine** (though after this film it was retconned such that Tozzi wasn't playing Frost). Alice Eve was close to signing, but negotiations broke down late in the day. It is possible that Oliver Platt's character was intended to be Fred Duncan, an FBI agent from the early *X-Men* comics, but it is never confirmed. The film was originally touted as a reboot of the series, but for the most part the film is fairly faithful to the overall continuity of the series.

**Comic Notes:** A lot of the characters in this movie have appeared in previous films – even if only very briefly – so take a look back at the earlier entries for those characters.

*Emma Frost*: first appeared in *Uncanny X-Men #129 (Jan, '80)*, created by the incomparable Chris Claremont and John Byrne. She is a major character in the comics, and has a great deal of character background that the movie version clearly doesn't. In the comics, she is the daughter of a wealthy couple, and has an older brother Christian, an older sister Adrienne and a younger sister Cordelia – all of whom have appeared at some point in the comics. Initially nothing more than a telepath, she left her family and became involved with crooks. As her powers grew she was noticed by the Hellfire Club which she became part of, and then alongside Sebastian Shaw she became the White Queen. She also set up the Massachusetts Academy for mutants – a rival to Xavier's school – which even had its own form of X-Men, the Hellions. She supported Magneto becoming White King of the Club, and supported him when he ousted Shaw. When the Hellions were killed, she was extremely traumatised by the events, and thanks to Xavier she overcame them, and joined the X-Men, even becoming co-leader with Scott Summers, as well as his lover. Whilst on Genosha, another mutant, Cassandra Nova, forced a second mutation on her, giving her a diamond form which was essentially invulnerable, in order to survive Genosha's destruction. She currently works with Scott, Jean, and her clones, the Stepford Cuckoos as part of the X-Men.

*Sebastian Shaw*: who has the same powers his movie counterpart has, made his appearance at the same time as Emma did. Unlike his movie counterpart, he was not a Mengele-style character during World War II. Instead he was the billionaire head of Shaw Industries, invited to join the Hellfire Club by Ned Buckman as the Black Bishop. Shaw became Black King of the Club, alongside Donald Pierce (the White King), Emma Frost and Selene (the Black Queen). He was constantly attended by a woman named Tessa. Ironically the Lords Cardinal took control of the club after defeating the Sentinels, and Shaw then supported Robert Kelly, Henry Peter Gyrich and Boliver Trask in creating more Sentinels, appearing to them as an anti-mutant campaigner. The Club fought the X-Men, though they joined forces when a future Sentinel named Nimrod attacked, and with Pierce gone, Magneto took the mantle of White King. Since then Shaw has gained and lost his Black King title (sometimes taken from him by his son Shinobi). Most recently Shaw had his memory wiped by Emma, got it back from an Avengers file, and rescued a group of mutant children, leaving his status somewhat grey, rather than black.

*Alex Summers*: has the same abilities as his movie counterpart, but his history is rather interesting. In the movies it's unclear if Alex and Scott are related, or coincidentally have the same surname; in the comics, Alex is definitely Scott's younger brother (unless you're reading *Ultimate X-Men* where he's the older sibling). He first appeared in *The Uncanny X-Men #54 (Mar, '69)*, created by Arnold Drake, Don Heck and Neal Adams. Alex, like Scott, survived the plane crash they were in, and was adopted out quite quickly. His powers manifested when he incinerated a boy who caused the death of his foster brother, but Mr Sinister blocked his powers. While studying geophysics he met the X-Men and his brother. Unable to control his powers exactly, he didn't immediately join the X-Men, but acted as a reserve member, though later did join, and started a long term on/off relationship with Lorna Dane (or Polaris).

The pair went back to being reserve members, though Havok re-joined when Lorna was taken over by Malice, and he ended up meeting Madelyne Pryor, his sister-in-law, who manipulated him and the X-Men into fighting Scott Summers' X-Factor. Like his brother, he has died and come back – though in fairness, it was only apparent as he was flung into a parallel universe – and currently stands with is brother's renegade mutant team.

*Sean Cassidy*: or Banshee, first appeared in *The X-Men #28 (Jan, '67)* created by Roy Thomas and Werner Roth. Like the movie version he has the sonic scream, which can give him flight, but he also has superhuman hearing. He is the heir to Cassidy Keep and the fortune that goes with it, and he married Maeve Rourke with whom he had a daughter, Teresa (see **X2** for more on her). His cousin, the so-called Black Tom Cassidy kept his daughter a secret from him, as Cassidy went to work for Interpol, and then became a private investigator. He briefly encountered the X-Men, and was later recruited to join Xavier's second team, which he accepted. Most interestingly, while learning of his daughter and recovering from a fight which almost cost him his powers, he fell in love with Moira MacTaggert. Unsurprisingly, he lost his life while trying to stop Gabriel Summers, but was resurrected by the Celestial Death Seed, and made a Horseman by the Apocalypse Twins. He currently remains in Avengers custody until Beast can cure him of the influence of Death Seed and the Twins.

*Armando Muñoz*: or Darwin, first appeared in *X-Men: Deadly Genesis #2 (Feb, '06)*, created by Ed Brubaker and Pete Woods, and does indeed have the power of retroactive evolution. The comic version was found by Moira MacTaggert, and he stayed with her and became friends with Gabriel Summers – the third and youngest of the Summers brothers. He has been recruited to the X-Men by Xavier on occasion, and is currently working with X-Factor.

*Angel Salvadore*: first appeared in *New X-Men #118 (Nov, '01)*, created by Grant Morrison and Ethan Van Sciver. Like the movie counterpart she has wings – though sadly, they appeared after she formed a cocoon – and Wolverine then recruited her to Xavier's school. She could also spit acid, though since the House of M incident, she lost all her powers. Before that, however, she and Beak – a mutant who looks a bit like a chicken – became sexually involved and thanks to her insect physiology, Angel laid eggs quite soon, giving birth to a number of very curious children. Beak and the children also lost their powers, but both she and Beak now use technological powers to fight their cause; Angel going under the name Tempest.

*Azazel*: first appeared in *Uncanny X-Men #428 (Oct, '03)*, created by Chuck Austen and Sean Phillips. Whilst the movie version is essentially a teleporting mutant, the comic version is very different (though not appearance wise). Azazel claims he hails from Biblical times and is the leader of the demonic mutant horde called the Neyaphem who were banished to an alternate dimension by the Cheyarafim. Thanks to his teleportation powers he can occasionally return to Earth where he hopes to impregnate someone allowing him a permanent link to Earth. He targeted Mystique, with whom he fell in mutual love, and she gave birth to Kurt Wagner – though Azazel left in order to prevent her being a target of the Cheyarafim (not that she knew this and as such, bitterly hated him because of it). Mephisto and Blackheart have both suggested that Azazel is in fact a demon, or at least has demon blood. This may explain why he also possesses the ability to shapeshift and use energy blasts, alongside his immortality, telepathy and healing factor.

*Janos Quested*: (Riptide) first appeared in *The Uncanny X-Men #210 (Oct, '86)*, created by Claremont, John Romita Jr and Dan Green as a member of Mr Sinister's Marauders, dispatched to kill the Morlocks in the Mutant Massacre. His powers are much the same as the movie version's, but not enough to fell Colossus, who killed him. He was cloned by Sinister, and returned during Inferno (a story arc in which the demon N'astirh attempted to use his demon horde to escape Limbo and conquer the Earth using Madelyne Pryor and her child), only to be killed again. More Riptides have been seen since then.

**Ratings:** *IMDB:* 78%; *Rotten Tomatoes:* 87%; *Metacritic:* 65

**Review:** A film of two halves, the first being the X-Men side of things which is done brilliantly. McAvoy and Fassbender are perfect younger versions of the characters we've already seen, and to see them forging a team together against a sixties backdrop that is more **Austin Powers** than reality is fantastic. However, Sebastian Shaw's Hellfire Club seems like a cheap version of the Brotherhood of Mutants, and Shaw never really gives any sense of menace, which weakens the movie a little. However, the X-Men side is so good, it makes the whole thing entertaining.

## GHOST RIDER: SPIRIT OF VENGEANCE
### *(17/2/2012)*

Moreau, a priest, warns his monastery of an attack to get hold of a boy named Danny that is wanted by Mephistopheles, and as the attack takes place, Moreau spirits the boy and his mother away, seeking out the help of Johnny Blaze, who has lain low for some time, fearful of the indiscriminate power of the Ghost Rider. Danny, and his mother Nadya, find themselves captured by her former boyfriend Ray Carrigan, but they are rescued by Ghost Rider, and the three find themselves together on the run from Carrigan and his boss Roarke – in actuality Mephistopheles himself.

**Cast:** *Nicholas Cage (Johnny Blaze/Ghost Rider),* Violante Placido (Nadya), *Ciarán Hinds (Roarke),* Idris Elba (Moreau), *Johnny Whitworth (Ray Carrigan), Fergus Riordan (Danny),* Spencer Wilding (Grannik), Sorin Tofan (Kurdish), Jacek Koman (Terrokov), Anthony Head (Benedict), Christian Iacob (Vasil), Christopher Lambert (Methodius), Jai Stefan (Krakchev), Vincent Regan (Toma Nikasevic), Ionut Christian Lefter (Young Johnny Blaze)

**Dir:** Neveldine/Taylor; **Writers:** David S Goyer, Scott M Gimple, Seth Hoffman; **Prod:** Avi Arad, Michael de Luca, Ashok Amritraj, Steven Paul; **Music:** David Sardy; **Exec.Prod:** Gary Foster, Maya Gallagher, David S Goyer, E Bennett Walsh, Mark Steven Johnson, Stan Lee; **DOP:** Brandon Trost; **Prod.Des.:** Kevin Phipps; **Ed:** Brian Berdan; **Costumes:** Sammy Sheldon; **Crystal Sky Pictures/Hyde Park Entertainment/Imagenation Abu Dhabi/Marvel Entertainment**; 95; c $65m; $132m

**Notes:** Directors Mark Neveldine and Brian Taylor cameo in the film, shaking hands as Johnny talks of making a deal with the devil. Though not credited as such, Ciarán Hinds is playing Mephistopheles, the same character Peter Fonda portrayed in the earlier film, and one Fonda hoped to reprise. The rights to Ghost Rider have now reverted back to Marvel Studios. According to Neveldine and Taylor, the filming was troubled as despite establishing locations in Europe to film, these locations would be unavailable when filming commenced, necessitating some last minute changes.

**Comic Notes:** *Ray Carrigan:* once transformed by Mephistopheles, actually becomes the comic villain Blackout, though he is never named as such. Blackout first appeared in *Ghost Rider #2 (Jun, '90)* [the 1990 run was Vol. 3], and was created by Howard Mackie and Javier Saltares. He could certainly extinguish light, similar to the movie version, though he had no decay power; rather he had super human strength and metal fingernails. The comic version is also a demonic hybrid, but the comic version has little else in common with the movie version.

    *Danny:* is intended to be Danny Ketch. He was the second Ghost Rider, appearing for the first time in *Ghost Rider #1 (May, '90),* also created by Mackie and Saltares. Ketch became the Ghost Rider after he and his sister were attacked by thugs, and he touched a motorcycle with a magical sigil on it. Later he learned that he was Johnny Blaze's half-brother – they were both the sons of Barton Blaze. Given the situation in the movie, it's unlikely that this is the case here.

**Ratings:** *IMDB:* 43%; *Rotten Tomatoes:* 17%; *Metacritic:* 32

**Review:** In the previous version of this book I was quite positive about this film, but on reviewing it again for this volume, I realise I was far too generous. The movie has a few nice moments – such as turning the giant mining equipment into Ghost Rider's vehicle – but there's not enough of these moments to genuinely entertain. More importantly, Nicholas Cage's acting seems ridiculously out of place in this film. It's a step up from the last one, but it's just too crazy this time round. More importantly, Ghost Rider loses the scariness he really should possess. It's not an effective movie, but worse, it's not entertaining.

## THE AMAZING SPIDER-MAN
### *(30/6/2012)*

As a child, something happens to Peter Parker's parents, and he is deposited with his aunt and uncle, and they disappear, to be killed in a plane crash. Years later, Peter is curious about his father's work, and this brings him to OsCorp where the man worked, and into contact with Dr Curt Connors. Accidentally bitten by a radioactive spider, Peter is endowed with super powers, and when he gives the missing piece of a formula to Connors, Connors concocts a serum to regenerate his missing arm with lizard DNA – though the effects on Connors are far worse than the ones on Peter.

**Cast:** *Andrew Garfield (Spider-Man/Peter Parker), Emma Stone (Gwen Stacy), Rhys Ifans (The Lizard/Dr Curt Connors), Denis Leary (Captain Stacy), Martin Sheen (Uncle Ben), Sally Field (Aunt May),* Irrfan Khan (Rajit Ratha), *Campbell Scott (Richard Parker), Embeth Davidtz (Mary Parker), Chris Zylka (Flash Thompson), Kelsey Chow (Hot Girl),* Michael Massee (Man in the Shadows)

**Dir:** Marc Webb; **Writers:** James Vanderbilt, Alvin Sargent, Steve Kloves; **Prod:** Laura Ziskin, Avi Arad, Matt Tolmach; **Music:** James Horner; **Exec.Prod:** Stan Lee, Kevin Feige, Michael Grillo; **DOP:** John Schwartzman; **Prod.Des.:** J Michael Riva; **Ed:** Alan Edward Ball, Michael McCusker, Pietro Scalia; **Costumes:** Kym Barrett; **Columbia Pictures/Marvel Entertainment**; 136; c $215m; $758m

**Notes:** Despite the not-so-positive reaction to **Spider-Man 3**, the decision was always that Raimi would return and proceed with a fourth film. Raimi had ideas for the Lizard to appear (having seeded him in two of the films) and also for the Vulture and the Black Cat (played by John Malkovich and Anne Hathaway). However, with a release date of May 5, 2011, the costs apparently getting too high and tensions forming between Arad and both Raimi and Maguire, Raimi felt he couldn't proceed with the film and decided to step down. In a surprise move, the producers decided to reboot the series, handing the reigns over to relatively new director Marc Webb. The writers decided to keep the Lizard as well as going to a more comic book based explanation for the web shooters. Michael Massee's character was set up to remain mysterious, though the next movie would give some more background on the character.

**Stan Spotting:** In what is arguably his best cameo ever, Stan plays the school librarian who, thanks to headphones, is blissfully unaware of the fight between Spider-Man and the Lizard that is going on behind him.

**Comic Notes:** Most of these characters appeared in previous films, so their notes can be found in those entries. However…Peter's parents, Richard and Mary, first appeared in *The Amazing Spider-Man Annual #5 (1968)* in flashback, and were created by Stan Lee and Larry Lieber. They both worked for the CIA, rather than OsCorp, and while on a mission to fight the Red Skull (the Russian Albert Malik, rather than the original Johann Schmidt) they were caught and killed. There does seem to be some question about their deaths, though, as throughout comic history, several villains have claimed that the Parkers were alive – though in truth, none of them could justify this. Curiously, in *Ultimate Spider-Man*, Richard was a biologist, and he and Mary both died in a plane crash while he was working on a suit that could regenerate the body – a suit that would ultimately be the basis for Venom. Dr Curt Connors continued their research after they died; all of which sounds a little more like the movie than the original comics do.

*Sally Avril:* Although credited as "hot girl", Kelsey Chow stated that her character was Sally Avril. Sally first appeared in *Amazing Fantasy #15 (Aug, '62)* and she was created by Lee, Steve Ditko, Kurt Busiek and Pat Olliffe. Lee and Ditko wrote her much as she appears in the movie – a student at Peter's school, who is more interested in Flash Thompson than Peter Parker. Busiek and Olliffe subsequently reinvented the character as a fantastic gymnast who, smitten by Spider-Man, would adopt a costume and become The Bluebird, a somewhat incompetent superhero. Rather brutally, Spider-Man allowed the Black Knight to severely beat her in the hope that she would be put off, but Sally was a surprisingly tough character. Also a photographer, she went to photograph Spider-Man's battle against the Knight, but was killed in a car crash on the way, leaving Peter Parker particularly remorseful.

**Ratings:** *IMDB: 71%; Rotten Tomatoes: 72%; Metacritic: 66*

**Review:** Reboots are not inherently bad things, but multiple origin stories are somewhat testing, and as such, for the first part of the film, the audience is left very much with a "been there, done that" sort of feeling. However, Andrew Garfield is probably a better choice for Peter Parker, and certainly brings Spider-Man to life a little better, while the chemistry between him and Stone easily surpasses Maguire and Dunst. The film is a lot of fun, and while it's perhaps not as good as Maguire's first two films, it's much better than his third and well worth watching.

## To Reboot Or Not To Reboot...

Film series have a curious lifespan, and the sort of films that follow the first can be broken down into several different categories.

A sequel, technically, is a film that carries on the story of the first one. There are surprisingly few genuine sequels in action/adventure films. The James Bond series, for instance, merely has a lot of films, none of which are really sequels (with the exception of **Quantum of Solace** or **SPECTRE**). Equally, most comic book movies are similar – the second film isn't really a sequel, it's simply another film using the same base set. Even **Blade II**, which although it sort of carries on the story from the first film, really just resolves it to move onto something more interesting. There are very few **The Godfather Part II**'s here.

A reboot, on the other hand, resets everything. It starts the entire process again, essentially tossing away what happened before it, and establishing a new continuity. That said, there are different ways of doing this. It's not necessary to show the origins of the characters in order to reboot something. You can just mention anything like that in passing, rather than having to waste time on something that the audience might be broadly familiar with. Both **The Amazing Spider-Man** and **Spider-Man: Homecoming** were reboots of the film series, but whereas the former was a new origin for Peter Parker, the latter assumed that the audience was familiar with how Peter Parker became Spider-Man (though they might not necessarily know the exact details) and the story can move forward from there.

Gale Anne Hurd also came up with a third variation of this – the requel. A requel is a movie that reboots the series, whilst also carrying on the story from the first film. **The Incredible Hulk** is the best example of this, with certain elements of the film clearly establishing a new continuity from **The Hulk**. Yet despite that, the movie does pick up from where the first film left off.

Which is the best option is hard to say, of course, and invariably is dependent on what the filmmakers wish to achieve. However, in truth, the idea of a requel is probably the best one that anyone has come up with; essentially allowing the filmmakers to do whatever they like, while at the same time leaving explanations up to the movies that came before it, if necessary.

## THE WOLVERINE
### *(16/7/2013)*

In 1945, Logan, a POW, is present when Nagasaki is bombed and as it happens he saves the life of Ichiro Yashida. In present day, haunted by his murder of Jean Grey, Logan attempts to live under the radar, but a girl named Yukio finds him and takes him to Yashida, who is dying. Shingen Yashida, Ichiro's son, wants to transplant Logan's healing factor to save Ichiro's life, and take the immortality curse away from Logan, but he refuses. However, at Yashida's funeral, gunmen attempt to kidnap the deceased's granddaughter Mariko, and Logan rescues her, leading to the two going on the run from the Yakuza and the search for what they want.

**Cast:** *Hugh Jackman (Logan), Tao Okamoto (Mariko), Rila Fukushima (Yukio), Hiroyuki Sanada (Shingen), Svetlana Khodchenkova (Viper),* Brian Tee (Noboru), Hal Yamanouchi (Yashida), *Will Yun Lee (Harada), Famke Janssen (Jean Grey)*

**Dir:** James Mangold; **Writers:** Mark Bomback, Scott Frank; **Prod:** Lauren Schuler Donner, Hutch Parker; **Music:** Marco Beltrami; **Exec.Prod:** Stan Lee, Joe Caracciolo Jr; **DOP:** Ross Emery; **Prod.Des.:** François Audouy; **Ed:** Michael McCusker; **Costumes:** Isis Mussenden; **20th Century Fox/Donners' Company/Marvel Entertainment/TSG Entertainment/Ingenious Media/Big Screen Productions**; 126; c $115m; $415m

**Notes:** Both Patrick Stewart and Ian McKellan make uncredited cameos in the mid-credits sequence as Xavier and Magneto, leading into **Days of Future Past**. The movie is based loosely on Chris Claremont & Frank Miller's *Wolverine* series, in which the Japanese link to Logan was first made. An alternate ending for the film saw Yukio giving Wolverine his traditional comic book mask, though this was ultimately dropped. Similarly to **First Class**, Jessica Biel was close to being cast as Viper, until the deal fell through for some reason. Darren Aronofsky was announced as the director of the film, but five months later he pulled out, unwilling to be out of the country for the length of filming. Hugh Jackman is an uncredited producer of the film.

**Should I Stay To The End?** Oddly, there is a mid-credits sequence where Wolverine encounters both Magneto and the surprisingly alive Xavier who need his help, so yes, stay.

**Comic Notes:** As mentioned this is based on Claremont and Miller's run of *Wolverine (Vol 1)*, and as such there are a few characters appearing, though surprisingly the movie does make a fair few changes to the characters.

*Mariko Yashida*: made her first appearance in *The Uncanny X-Men #118 (Feb, '79)* created by Claremont and John Byrne. Mariko is much the same as her movie character – the daughter of Shingen – though she was forced to marry a crime boss who beat her, which Wolverine dealt with, and from this their love became an engagement. She was poisoned with tetrodotoxin, and asked Wolverine to end her life, rather than die a drawn out death, which he agreed to.

*Lord Shingen Yashida and Yukio*: both debuted in *Wolverine #1 (Sep, '82)*, and of course were created by Claremont and Miller. The former is somewhat similar to his movie counterpart; Wolverine confronting him because of his treatment of Mariko. Shingen had Wolverine poisoned, and when they duelled he won, though they would fight again and Wolverine would kill the man. Although Wolverine decimated the Yashida organisation, its remains were taken over by Mariko, similar to the movie. Yukio is radically different to her movie persona, not least in appearance – in the comics she has short black hair and wears a leather body suit, nothing like the long red haired, anime-inspired character of the movie. Equally, the comic version is not a mutant, simply an assassin working for Shingen (indeed, it was she who poisoned Wolverine before his duel with Shingen). Latterly, at Logan and Mariko's wedding (which never came to pass), Viper and Silver Samurai attempted to attack the X-Men, and Yukio joined them in fighting the pair. She later became a source of information for Charles Xavier. When Wolverine was possessed, becoming Hellverine, he badly wounded Yukio, confining her to a wheelchair for the rest of her life. As an aside, Wolverine once saved a young girl – Amiko Kobayashi – and left her with Mariko to mother. When Mariko died, Amiko was briefly fostered out, before Yukio became her guardian, and Amiko currently remains by her side.

*The Silver Samurai:* The character of Harada is quite different to his comic version as well. In fact, in the comics it is not Ichiro Yashida who is the Silver Samurai, it is Kenuichio Harada; and he is not Mariko's former lover, but her half-brother. He first appeared in *Daredevil #111 (Jul, '74)* created by Steve Gerber and Bob Brown, and often fought alongside Viper. Harada is actually a mutant, able to create a tachyon field around anything; most often around his katana allowing it to slice through almost anything. He has had a spotted history, starting as a villain and assuming control of the Yashida clan when Mariko died, but he later attempted to redeem his clan and worked alongside Wolverine. At one point he was even the leader of Big Hero 6, the first Japanese superhero team. He was killed by ninjas working for the Red Right Hand, and the Silver Samurai became his son, Shingen Harada. He was ostensibly also a mutant, joining the Brotherhood, though what his power is remains unclear.

*Viper:* The character of Viper is also significantly changed from the comics. In the comics she has no powers, but is merely a highly skilled and trained athlete, her real name being Ophelia Sarkissian. She first appeared in *Captain America #110 (Feb, '69)* created by Jim Steranko, under the alias Madame Hydra where she was working for...well, Hydra. She'd been orphaned and then taken by Hydra and trained by Kraken. She would later leave Hydra and join forces with Viper, the head of the Serpent Squad, until she turned on him and took control of the Squad as the new Viper. She has had a surprisingly busy career since then – almost always criminal – including working with (and possibly also the lover of) the Silver Samurai, attempting to seize control of the Serpent Society, keeping a group of clones called the Pit-Vipers, blackmailing Wolverine into marriage, joining the Hellfire Club and becoming dictator of Madripoor, where she currently appears to be.

**Ratings:** *IMDB:* 67%; *Rotten Tomatoes:* 70%; *Metacritic:* 60

**Review:** Hugh Jackman was apparently very keen to "do Wolverine right" after disappointing results in the previous film, and this time round the film does indeed hit the target a little better. Rooting the movie in Frank Miller's comic run was a good start, and gives Wolverine the perfect opportunity to confront new villains, and make new allies, as well as fall in love. What doesn't work quite so well is Wolverine's obsession with Jean Grey, and the fact that the Mariko love affair is all but forgotten by the end of the film. Extremely enjoyable, but still not quite the hit Jackman was hoping for.

## THE AMAZING SPIDER-MAN 2
### *(10/4/2014)*

A brief encounter with Spider-Man changes Max Dillon's life, and he becomes moderately obsessed with the superhero, believing himself to be his best friend. Peter rekindles his relationship with Gwen while still remaining Spider-Man, though he feels guilty about not obeying Captain Stacy's dying wish, which causes a number of hurdles with Gwen. Meanwhile Harry Osborn has re-entered Peter's life, though Osborn has his own problems as his father's company is stolen from him. When Dillon accidentally gains electrokinesis, his life changes, affecting Peter's, Gwen's and Harry's.

**Cast:** *Andrew Garfield (Spider-Man/Peter Parker), Emma Stone (Gwen Stacy), Jamie Foxx (Electro/Max Dillon), Dane DeHaan (Green Goblin/Harry Osborn), Colm Feore (Donald Menken), Felicity Jones (Felicia), Paul Giamatti (Aleksei Sytsevich), Sally Field (Aunt May), Campbell Scott (Richard Parker), Embeth Davidtz (Mary Parker), Marton Csokas (Dr Ashley Kafka), B J Novak (Alistair Smythe),* Michael Massee (Gustav Fiers/The Gentleman), *Chris Cooper (Norman Osborn), Denis Leary (George Stacy)*

**Dir:** Marc Webb; **Writers:** Alex Kurtzman, Roberto Orci, Jeff Pinkner, James Vanderbilt; **Prod:** Avi Arad, Matt Tolmach; **Music:** Hans Zimmer, The Magnificent Six; **Exec.Prod:** Alex Kurtzman, Roberto Orci, E Bennett Walsh, Stan Lee; **DOP:** Dan Mindel; **Prod.Des.:** Mark Friedberg; **Ed:** Pietro Scalia; **Costumes:** Deborah Lynn Scott; **Columbia Pictures/Marvel Entertainment**; 142; c $245m; $709m

**Notes:** Spider-Man's costume is changed for this film in order to return it more to how it was in the comics. When Gwen dies, it is 121 minutes into the film, and a clock shows the time as 1.21. In the comics, Gwen dies in *The Amazing Spider-Man #121.* Michael Massee's character is revealed to be Gustav Fiers. Fiers was a wealthy industrialist who assembled the Sinister Six in the 1999 novel *Spider-Man: The Gathering of the Sinister Six.* Marc Webb was actually contracted to Fox at the time of this production, with another movie left on his contract. In order to be allowed to do this film, Webb had to extend his contract with Fox by another film, and Sony were required to promote **X-Men: Days of Future Past**. As such, unusually, the theatrical version has a mid-credits sequence in which the trailer for the Fox film is shown (the DVD release does not have the trailer). The original outline for the movie changed somewhat dramatically as the film went into production, with a number of elements being dropped in favour of bringing certain others to the fore. Oddly, the character of Felicia was cut considerably; originally she was more clearly defined as Harry Osborn's girlfriend, helping him throughout the film. Dr Ratha was supposed to return but Irrfan Khan was unavailable and the character was replaced with Colm Feore's Donald Menken. Richard Parker would have turned up at the end of the film, uttering the famous "with great power comes great responsibility" line, and Peter would voluntarily give his blood to Harry. However, the largest cut was Mary Jane Watson. Shailene Woodley was cast as Mary Jane, but despite filming her scenes (and indeed, Woodley can be seen in the film if you look carefully enough at the graduation scene), it was decided that, given the ending of the film, introducing Mary Jane would be too much, and as such the character was cut. The original intention after this film was for two more Amazing Spider-Man films, a Sinister Six film (to be written and directed by Drew Goddard), a Venom film (written by Kurtzman, Orci and Ed Solomon and directed by Kurtzman) and a film centred around a female protagonist (presumably the Black Cat). However, certain executives were unhappy with the film's reception and with Andrew Garfield. Additionally, while the film was enormously successful, the box office results weren't as great as the previous film, which was cheaper to make. Sam Raimi was approached in regard to returning to the series, while talks were held with Marvel in the hopes of pushing Spider-Man into the MCU. However, Marvel were unhappy with Sony's terms and turned them down. The fourth Amazing Spider-Man film was cancelled, while the third was pushed back, and the Sinister Six film was brought forward. It was then that the infamous Sony e-mail leak occurred, exposing all of the behind the scenes problems to, not only the public, but also to Marc Webb and Andrew Garfield. With studio executive Amy Pascal's job on the line (for a large variety of reasons), she returned to Marvel with a new offer, and an announcement was made which saw Garfield and Webb effectively fired, and a new Spider-Man film to be co-produced by Marvel Studios. (Pascal, incidentally, would also be fired for racist comments about then-President Barack Obama, but would be given movie production roles as a payoff).

**Stan Spotting:** One of the guests at Peter and Gwen's graduation.

**Comic Notes:** *Max Dillon*: or Electro as he is better known, first appeared in *The Amazing Spider-Man #9 (Feb, '64)*, created by Stan Lee and Steve Ditko. He was an electrical engineer in the comics, who was hit by lightning, which caused a mutation that turned him into, effectively, a capacitor. Though he has often fought many Marvel superheroes, his arch nemesis is Spider-Man, and he is often a member of the Sinister Six. His original appearance was a green costume with yellow lightning mask (which is nodded to on Dillon's birthday cake in the movie), but the movie version is based more on the cartoon version of Electro. Foxx's casting is against type for the character, who is not African American.

    *Aleksei Systevich*: The underused Rhino, first appeared in *The Amazing Spider-Man #41 (Oct, '66)*, created by Lee and John Romita Sr. Rhino does have armour, though it is more of a bodysuit in the comics than a giant metal rhinoceros, but Systevich was actually experimented on in the comics, and as such has superhuman strength, speed and stamina, which is very different to the version seen in this film. He too has been a frequent member of the Sinister Six.

*Felicia Hardy:* A marketing video for the film confirmed that Felicia's last name is Hardy, making her the Black Cat. Felicia first appeared in *The Amazing Spider-Man #194 (Jul, '79)*, and was created by Marv Wolfman and Keith Pollard. She is the daughter of a cat-burglar, who trained herself to follow in her father's footsteps. She studied a number of fighting styles after she was raped at university, and she became the Black Cat to break her father out of prison (though ironically he died before she was able to). The Black Cat and Spider-Man developed a romantic relationship over time, though Felicia struggled to accept that Peter Parker was Spider-Man. She is, of course, nothing like DC's Catwoman. They are not remotely similar. In any way. Clearly…

*Donald Menken:* first appeared in *The Amazing Spider-Man #239 (Apr, '83)*, created by Roger Stern and John Romita Jr. He was Norman Osborn's personal assistant, but was a largely background character.

*Dr Ashley Kafka:* was a professional psychologist specialising in the criminally insane, who worked at Ravencroft Institute for some time. As such, she is very similar to her movie counterpart, with the exception that the movie version is male. Sadly Dr Kafka was killed by Massacre, who ripped her eyes out of her head to get past a retinal scanner. She first appeared in *The Spectacular Spider-Man #178 (Jul, '91)* and was created by J M DeMatteis and Sal Buscema.

*Alistair Smythe:* first appeared in *The Amazing Spider-Man Annual #19 (Nov, '85)*, created by Louise Simonson and Mary Wilshire. He is radically different to his movie counterpart. In the comics, Alistair is the son of Spencer who created the Spider Slayer robots that J Jonah Jameson funded in order to kill Spider-Man. Alistair vowed to avenge his father, who died because of the creation of the robots, and went to work for the Kingpin to bring the wallcrawler down. After spending time in prison, Alistair built a bioorganic shell and dubbed himself the Ultimate Spider-Slayer and would continue to fight Spider-Man. He was killed by Spider-Man, though ironically, it was when Peter Parker's body was inhabited by Otto Octavius's mind.

**Ratings:** *IMDB:* 63%; *Rotten Tomatoes:* 53%; *Metacritic:* 49

**Review:** After a strong restart to the series, pretty much everything goes tails up in this film. No matter who good Garfield and Stone's casting maybe, virtually every other person in this film is miscast. Jamie Foxx, Paul Giamatti and Dane DeHaan are all terrible for different reasons; Giamatti is wasted, DeHaan is ridiculous as the Green Goblin, and woeful compared to James Franco as Harry Osborn, while Jamie Foxx struggles to nail his character in any sensible way. Add to that Hans Zimmer annoying score, plotlines about the Parkers that ultimately go nowhere, and the fact this film is clearly setting up a number of films to follow it, and you have a film that has no real identity. It's a shame because the last twenty minutes of the film are perfect. Unfortunately, getting there is far too difficult.

## X-MEN: DAYS OF FUTURE PAST
### (10/5/2014)

Xavier and Magneto formulate a plan to alter history and prevent the construction of the Sentinels who currently monitor, capture, imprison and execute mutants. The plan depends on Kitty Pryde sending Wolverine's mind back to the 1970s. Once there he attempts to locate Xavier and Magneto; only to find the former is a recluse, addicted to a mutant-hampering drug; and the latter is in jail for assassinating Kennedy. All are needed to locate Mystique and stop her from executing Boliver Trask; an action which will inadvertently lead to the creation of the perfect Sentinels.

**Cast:** *Hugh Jackman (Logan/Wolverine), James McAvoy & Patrick Stewart (Charles Xavier), Michael Fassbender & Ian McKellan (Erik Lehnsherr/ Magneto), Jennifer Lawrence (Raven Darkholme/Mystique), Nicholas Hoult & Kelsey Grammer (Hank McCoy/Beast), Anna Paquin (Rogue), Ellen Page (Kitty Pryde/Shadowcat), Peter Dinklage (Boliver Trask), Shawn Ashmore (Bobby Drake/Iceman), Omar Sy (Bishop), Evan Peters (Peter Maximoff/ Quicksilver), Josh Helman (William Stryker), Daniel Cudmore (Peter Rasputin/ Colossus), Fan Bingbing (Blink), Adan Canto (Sunspot), Booboo Stewart (James Proudstar/Warpath), Famke Janssen (Jean Grey), James Marsden (Scott Summers), Lucas Till (Alex Summers/Havok), Evan Jonigkeit (Toad), Gregg Lowe (Ink),* Mark Camacho (President Richard Nixon), Michael Lerner (Senator Brickman)

**Dir:** Bryan Singer; **Writers:** Simon Kinberg, Jane Goldman, Matthew Vaughn; **Prod:** Simon Kinberg, Hutch Parker, Lauren Schuler Donner, Bryan Singer; **Music:** John Ottman; **Exec.Prod:** Todd Hallowell, Josh McLaglen, Stan Lee; **DOP:** Newton Thomas Sigel; **Prod.Des.:** John Myhre; **Ed:** Michael Louis Hill, John Ottman; **Costumes:** Louise Mingenbach; **20th Century Fox/TSG Entertainment/Bad Hat Harry Productions/Donners' Company/Marvel Entertainment**; 132; $200m; $746m

**Notes:** Chris Claremont and Len Wein cameo as US Congressman. In earlier versions of the script, due to uncertainty regarding Hugh Jackman's involvement, Bishop was originally going to be the time traveller. Josh Helman was initially cast as Juggernaut, but this was changed to William Stryker, marking Helman as the third actor to play Stryker in as many films. Footage from all previous X-films is used at various points in the film. A later version of the film was released, dubbed "The Rogue Cut". This reinstated deleted scenes, including restoring most of Rogue's screen time, which in the theatrical version ultimately amounted to nothing more than a cameo. Although it is mostly deleted footage, the Rogue storyline necessitates the use of alternate scenes for some story elements.

**Comic Notes:** *Lucas Bishop:* first appeared in *Uncanny X-Men #282 (Nov, '91)* and was created by Whilce Portacio and Jim Lee. His powers are similar to his on-screen appearance, but the comic character is quite interesting: he actually comes from a dystopian future, and the "M" over his eye marks him as a mutant; as do others from his time. He travels back to present day and joins the X-Men, though over time has become more an adversary to them. Curiously, he is also Australian, unlike the movie where he is clearly French (at least in the Rogue cut it is clear).

*Clarice Ferguson:* Blink's first appearance was *Uncanny X-Men #317 (Oct, '94)* and she was created by Scott Lobdell and Joe Madureira. Her powers are also teleportation, though she doesn't appear to throw anything to make her portals. In the comics she has the same facial markings, but has purple skin and green eyes as well.

*Sunspot:* In the comics, his real name is Roberto DaCosta and he first appeared in *Marvel Graphic Novel #4: The New Mutants (Sep, '82)*. Created by Chris Claremont and Bob McLeod, Sunspot's powers are identical to the movie version.

*James Proudstar:* first appeared in *New Mutants #16 (Jun, '84)* as an enemy of the New Mutants, and using the identity Thunderbird, which was the codename his older brother used as an X-Man. He would latterly join the X-Men as Warpath. Warpath's powers seem similar to the movie version, though it's not entirely clear what they are in the film – effectively a perfect hunter. In the comics he can also fly and regenerate.

*Ink:* (Eric Gitter) first appeared in *Young X-Men #1 (Apr, '08)*, created by Marc Guggenheim and Yanick Paquette. In the comics his powers are many, and rely on the tattoos he has, though he most definitely has what we see in the movie: the biohazard tattoo giving the ability to make people ill.

*Bolivar Trask:* first appeared in *X-Men #14 (Nov, '65)* and was created by the Stan Lee/Jack Kirby team. In the comics he created the Sentinels, though he lost control of them to the Master Mold Sentinel. Similarly to Stryker he has mutant children. Trask appears to be in **X-Men: The Last**

**Stand**, and although that was supposed to be *the* Boliver Trask, Singer retconned this and suggested that the earlier version just happens to have a similar name.

*And:* Perhaps the most important comic note to make here, however, is that the film is based on *The Uncanny X-Men #141 - #142*, written by Chris Claremont, with art by John Byrne. In the comics, Rachel Summers (daughter of Scott Summers and Jean Grey in 2013), sends Kitty Pryde's mind back to 1980 where she and Wolverine attempt to stop Mystique's assassination of Senator Robert Kelly – an event which would lead to the mass use of Sentinels to hunt and imprison mutants.

**Ratings:** *IMDB:* 81%; *Rotten Tomatoes:* 91%; *Metacritic:* 74

**Review:** Singer returns and is on top form, giving us a brilliant film that manages to retcon everything before without being offensive or gratuitous about it. In many ways it's a Wolverine film, though the Xavier/Magneto/Mystique trio also make for an important element. Well-acted, and very well directed, the film is close to perfect. The Rogue cut adds some nice moments, though ironically, the Rogue sections are the least interesting and developed.

## The Fox X-Men Timeline

Thanks to a variety of directors, Bryan Singer's initial vision for the X-Men timeline is somewhat complicated, and from the outside doesn't actually appear to hold together. Surprisingly, however, it's not that contradictory, and thanks to the fact a number of movies merely give generalities in terms of dating (eg "The Vietnam War" rather than a specific year between 1955 and 1975) it is possible to move the dates around a little to make the timeline work. **Days of Future Past** has created a second timeline which has overwritten the events of some movies, and according to **Apoclaypse**, had a retro-effect on the timeline – the original timeline is designated **T-1**, the post **Days**, **T-2**. I speculate that the Sentinel timeline is a third timeline, **T-S**. Some events presumably take place in both timelines.

### c 2550 BC

En Sabah Nur builds a pyramid single handedly, while horsemen look on. (It's not clear what pyramid is being built, but assuming it's Giza's, then this date is based on when the actual pyramid was built.)

### 1845 AD

James Howlett watches his father killed by Thomas Logan. This activates his mutant powers and he kills Logan - only to discover Logan was his biological father. He and Logan's other son - Victor Creed - flee. (Date is established in **X-Men Origins: Wolverine.**)

## c 1865 AD

James Logan (formerly Howlett) and Creed fight in the American Civil War.

## c 1915 AD

Logan and Creed fight in the Great War.

## c 1930 AD

Erik Lehnsherr is born.

## c 1930 AD:T-2

Charles Xavier and Raven Darkholme are born.


Charles Xavier is born.

## 1944 AD

Dr Klaus Schmidt kills Erik Lehnsherr's mother, an action which creates the impetus for Lehnsherr to escape. Charles Xavier meets Raven, a runaway. (Date is established in both **X-Men** and **X-Men: First Class.**)

## 1945 AD

Wolverine is in a POW camp in Nagasaki when it is bombed. He saves Ichiroo Yashida. (Date is established in **The Wolverine.**)

## 1962 AD

Lehnsherr hunts Schmidt who now goes by the name Sebastian Shaw, and is the target of a CIA operation involving Agent Moira MacTaggert. A series of events sees Xavier - an Oxford University graduate - join forces with Lehnsherr to stop Shaw, who tries to take advantage of the Cuban missile crisis. While Shaw is stopped, Lehnsherr leaves with a group of followers, including Raven (now calling herself Mystique), Emma Frost and Azazel (both supporters of Shaw). Xavier is badly wounded and loses the ability to walk. His supporters include Alex Summers, Angel Salvadore and Hank McCoy. Wolverine appears not to be actively involved in the Vietnam War at this point. (Date is established in **X-Men: First Class.**)

## c 1963 AD:T-2

Jean Grey and Scott Summers are born. (Sophie Turner and Tye Sheridan were both born in 1996, which would have made them 20 during the filming of **X-Men: Apocalypse** which seems a reasonable assumption for the ages of Jean and Scott during that movie.)

Jean Grey and Scott Summers are born. (Famke Janssen was born in 1965 and Haley Ramm in 1992. Ramm was 14 when she filmed her scenes in **X-Men: The Last Stand**. Assuming Jean was also 14 in those scenes that would place her birth date here. This makes Jean seven years younger than Janssen. Similarly, James Marsden was born in 1973 and Tim Pocock in 1985. Marsden would be roughly the same age as Scott based on these dates. Pocock looked considerably older than 9 in **X-Men Origins: Wolverine**, though. Scott's age therefore becomes seemingly irreconcilable.)

## 1973 AD

Trask builds Sentinels. Wolverine, Beast, Magneto and Xavier stop Mystique killing Trask. Magneto escapes, Mystique pretends to be William Stryker and retrieves Wolverine. (Date is established in **X-Men: Days of Future Past**.)

## c 1974 AD

Events take place that result in Mystique giving up her impersonation of Stryker, and also losing custody of Wolverine. Stryker becomes an active member of the American Army as well.

## c 1975 AD

Stryker creates Team X, including Wolverine, Sabretooth, Agent Zero, Wade Wilson, Fred Dukes, John Wraith and Chris Bradley. (This takes place during the Vietnam War, but presumably is towards the end. Stryker as played by Danny Huston clearly looks older than the version played by Josh Helman, and so it presumably takes place after **Days of Future Past**.)

## 1975 AD:T-2

A car accident sees Jean Grey lose her mother, and her father happily gives her up to Charles Xavier, as he realises her actual power. (Date is established in **Dark Phoenix**.)

## c 1981 AD

Wolverine lives with Kayla Silverfox, who is murdered by Sabretooth. He agrees to undergo a process to have adamantium grafted to his skeleton. He escapes and joins forces with Remy LaBeau (Gambit) and John Wraith and they track Sabretooth to Three Mile Island where Stryker releases Deadpool. Silverfox (actually still alive), leads a number of mutants being held hostage by Stryker into the care of Charles Xavier. Kayla Silverfox's sister Emma has similar powers to Emma Frost. Wolverine is shot in the head which apparently causes some memory loss. (Takes place six years after Stryker's Team X.)

### 1983 AD:T-2
Magneto joins the newly awakened En Sabah Nur, who wishes to use Charles Xavier's powers to enslave mankind. Xavier, with his new X-Men: Jean Grey, Scott Summers, Kurt Wagner and Mystique fight Nur (Apocalypse) and his horsemen (Magneto, Storm, Angel, Psylocke). After his defeat, Mystique joins Xavier to train the X-Men, now including Storm. (And this single movie destroys the timeline. Date given in **X-Men: Apocalypse**.)

Xavier and Magneto meet Jean Grey for the first time. (Takes place twenty years prior to **X-Men: The Last Stand**.)

### 1992 AD:T-2
After the X-Men rescue a shuttle, Raven Darkholme feels the X-Men is straying from its ideals. However Jean Grey loses control and kills Raven, and Xavier admits he has placed mental blocks on Jean's mind. The D'Bari seek out the Phoenix entity which now resides in Jean, and their leader, Vuk, tries to convince Jean to give up the entity. When Jean realises how out of control she is, she confronts Vuk and sacrifices herself to destroy the D'Bari. (Date is established in **Dark Phoenix**.)

Warren Worthington II discovers his son has wings. (Takes place ten years prior to **X-Men: The Last Stand**.)

### c 2004 AD
Robert Kelly tries to pass the Mutant Registration Act, and Wolverine and Rogue become members of the X-Men, going up against Magneto's Brother of Mutants who want to kill Kelly. Wolverine has no memory of Xavier, and curiously Xavier shows no recognition of Wolverine. Sabretooth appears to have become more animalistic. Mystique has total loyalty to Magneto. (The movie is set in the near future. Placing it here means **X-Men: The Last Stand** is contemporaneous. This event seems to have happened in all timelines.)

## c 2005 AD

William Stryker attacks Xavier's school after Nightcrawler infiltrates the White House and attacks the President. The X-Men flee and reassemble, and confront Stryker at the Weapon X facility. The resultant confrontation causes the apparent death of Jean. The last mutant is born naturally. (Mystique is still impersonating Robert Kelly and Wolverine has been away for months, so it can't be too long after **X-Men**. The last mutant was born 25 years before **Logan**.)

Warren Worthington II announces the creation of a drug that can heal mutants, bringing Magneto's Brotherhood back into conflict with society and the X-Men coming up against them. Jean survived Alkali Lake, but kills Cyclops, and is ultimately killed by Wolverine. Xavier is also killed, though he appears to have survived by passing his consciousness into a coma patient, and contacts Moira MacTaggert - now a doctor. (Scott is still mourning Jean, so within a year of the previous film.)

## c 2007 AD

Yukio tracks down Wolverine - now a hermit - and takes him to Yashida in the hope he can save Yashida's life. Wolverine comes up against the Viper and Yashida, joining forces with Yukio and Mariko Yashida. (Takes place a year after Jean's death.)

## c 2010 AD:T-2

David Haller is broken out of his psychiatric ward and is forced to confront the Shadow King lurking in his mind. (An estimate for the date of **Legion**, based on the fact that Haller remembers his father's wheelchair as it appeared in **Apocalypse**, and the actor's age.)

## c 2013 AD:T-S

Wolverine – now with his adamantium claws restored - is approached by Xavier and Magneto to help confront a grave new threat. Presumably this is the new version of the Sentinels. (It's unclear how long after **The Wolverine** has taken place – I'm assuming this is contemporaneous with the film's release. This is, presumably, where the **Days of Future Past** future begins.)

## 2016 AD:T-2

Wade Wilson is diagnosed with terminal cancer. He decides to take a shady deal to cure him, but this results in kick starting his mutant abilities. Horribly disfigured, Wilson calls himself Deadpool and sets out to get his revenge on Ajax, the man that worked on him. (Presumably this is contemporaneous, though there is nothing to place it. Colossus is radically different to his previous appearances, and so that would suggest it doesn't take place in the same timeline as his other appearances – though it doesn't actively rule it out.)

### 2017 AD:T-2
The Strucker family are forced to go on the run from Sentinel Services, when their children turn out to be mutants. They join forces with the Mutant Underground; Reed Strucker offering to free Lorna Dane from prison in return for their help. (**The Gifted** is presumably contemporaneous.)


Laura Kinney is "born". (She is 11 in **Logan**.)

### 2023 AD:T-S
Mutants are hunted by Sentinels and imprisoned and/or killed. Small bands of mutants rebel, and Xavier's group join forces with Bishop's to implement a plan to alter history. (Date is established in **X-Men: Days of Future Past**.)

### 2023 AD:T-2
Wolverine returns to his body in a new timeline where Bobby and Rogue are still lovers, and Jean and Scott are still alive. (New future. Date is established in **X-Men: Days of Future Past**.)


The X-Men are killed by Charles Xavier as dementia causes him to lose control over his powers. (A year before **Logan**. New future.)


Logan and Caliban look after Xavier, though he makes contact with Laura, a clone of Logan. Racing against Zander Rice and his henchman Donald Pierce, Logan retrieves Laura and tries to get her to safety. (Date given in **Logan**. New future.)

# FANTASTIC FOUR
## *(4/8/2015)*

Reed Richards joins the Baxter Building, with the knowledge of how to bring something back after teleporting it away. Alongside Susan and Johnny Storm – the children of Franklin, the man in charge of the operation – and Victor Von Doom, they develop the device, and with Reed's friend Ben, take the trip to another world, which gives them bizarre powers. A trip which costs them Victor.

**Cast:** *Miles Teller (Reed Richards), Michael B Jordan (Johnny Storm), Kate Mara (Susan Storm), Jamie Bell (Ben Grimm/Thing), Toby Kebbell (Victor Von Doom/Dr Doom), Reg E Cathey (Dr Franklin Storm),* Tim Blake Nelson (Dr Allen), Dan Castellaneta (Mr Kenny), Owen Judge (Young Reed), Evan Hannemann (Young Ben), Chet Hanks (Jimmy Grimm)
**Dir:** Josh Trank; **Writers:** Simon Kinberg, Jeremy Slater, Josh Trank; **Prod:** Gregory Goodman, Simon Kinberg, Robert Kulzer, Hutch Parker, Matthew Vaughn; **Music:** Marco Beltrami, Philip Glass; **Exec.Prod:** Stan Lee; **DOP:** Matthew Jensen; **Prod.Des.:** Chris Seagers; **Ed:** Elliot Greenberg, Stephen E Rivkin; **Costumes:** George L Little; **20th Century Fox/TSG Entertainment/ Marv Films/Constantin Film/Marvel Entertainment/Genre Films/Moving Picture Company**; 100; c $140m; $168m

**Notes:** This movie had a particularly troubled development, with rumours abounding that Josh Trank's behaviour on set forced executives – led by Simon Kinberg – to step in, reshoot and ultimately edit the film. Indeed, on the day of release, Trank tweeted that he submitted a completely different version of the film a year earlier, which would have received better reviews; though the tweet was quickly deleted. A number of changes were made to the film – Dr Allen was originally Harvey Elder, the Mole Man; but this was changed as there wasn't a strong enough connection between the movie and the character. Toby Kebbell stated that his character was also changed, and rumour had it that the character was originally a computer hacker named Victor Domashev, who used the online name "Doom". When this was reported, Fox studios threatened legal action if they weren't taken down. In addition, most of the footage in the trailer does not make it to the movie. Much was made of Mara and Jordan's casting, with fans upset that they weren't biological siblings. Due to the reshoots, a number of continuity errors resulted; most notably Mara's hair, for which a wig was used the second time round. The film was originally to be converted to 3D, but this did not happen. A sequel was slated for 2017, but by November 2015, this was quietly dropped.

**Comic Notes:** Franklin Storm first appeared in *Fantastic Four #31 (Oct, '64)* and was created by Stan Lee and Jack Kirby. His backstory in the comics is completely different – a surgeon who turned to gambling after he lost his wife in a car accident, he ultimately found himself in jail after refusing to speak when placed on trial for the murder of a criminal in the employ of a loan shark (in truth that criminal killed himself). He was killed by a Skrull when trying to protect the Fantastic Four from a trap. This version of the Fantastic Four takes even more inspiration from *Ultimate Fantastic Four (Feb, '04)*, created by Mark Millar, Brian Michael Bendis and Adam Kubert. The stories are broadly similar – Reed was a social outcast who befriended the tougher Ben Grimm; Reed was accepted into Franklin Storm's group at the Baxter Building after a science fair; and alongside Victor Van Damme and Susan Storm, they created a bridge into N-Space, which gave Richards, Van Damme, Grimm, Susan and Johnny Storm their powers. Additionally, whereas the earlier films saw the Four more similar in age to their mainstream counterparts, this film has a younger team, like the *Ultimate* version. The movie differs significantly after the group build the teleporter, though.

**Ratings:** *IMDB:* 43%; *Rotten Tomatoes:* 9%; *Metacritic:* 27

**Review:** This is a hard film to be positive about. The first twenty minutes of the movie are bland, but after that the movie lurches from moment to moment, rarely with any real connective storyline. There is no chemistry between the four leads and no one seems to know exactly how to play their characters, which is only the first of many signs that there seems to be no direction. Doom's actions are completely nonsensical, hinted at only by a throwaway line towards the beginning of the film, but this is just one of many ideas that are hinted at, only to be ignored (the love triangle between Reed, Victor and Sue; the rebel that is Johnny, etc). It's a disastrous film; anything but fantastic.

## DEADPOOL
### *(12/2/2016)*

After spending a year of bliss with Vanessa, Wade Wilson discovers he has cancer and surrenders to a company for a radical treatment, which actually activates his mutant healing ability, whilst simultaneously scarring him horrifically. One year later, Wilson – now the vigilante Deadpool – brings Ajax to heel. Unfortunately, the X-Men are out to stop him.

**Cast:** *Ryan Reynolds (Wade Wilson/Deadpool), Morena Baccarin (Vanessa), Ed Skrein (Ajax), T J Miller (Weasel), Gina Carano (Angel Dust), Brianna Hildebrand (Negasonic Teenage Warhead), Stefan Kapicic (Colossus' Voice),*

*Greg LaSalle & Andre Tricoteux (Colossus' Movement Actors), Leslie Uggams (Blind Al)*, Karan Soni (Dopinder), Michael Benyaer (Warlord)

**Dir:** Tim Miller; **Writers:** Rhett Reese, Paul Wernick; **Prod:** Lauren Schuler Donner, Simon Kinberg, Ryan Reynolds; **Music:** Tom Holkenborg; **Exec.Prod:** John J Kelly, Jonathon Komack Martin, Rhett Reese, Aditya Sood, Paul Wernick, Stan Lee; **DOP:** Ken Seng; **Prod.Des.:** Sean Haworth; **Ed:** Julian Clarke; **Costumes:** Angus Strathie; **20th Century Fox/TSG Entertainment/ The Donners' Company/Kinberg Genre/Marvel Entertainment**; 108; $58m; $782m

**Stan Spotting:** The DJ at the strip club. Oh Stan...

**Notes:** Originally touted as one of several spinoff X-Men films, after the poor reception to the character in **X-Men Origins: Wolverine**, it was decided to pass on the film (though the poor reception was mostly due to the treatment of Deadpool, rather than the character itself). This wasn't actually the first attempt at a Deadpool film however – Artisan had the rights to the character and announced a film in 2000, which never came to pass. New Line Cinema acquired the rights and started development in 2004, but again the project stalled. In 2012, Tim Miller had three minutes of footage created, with Reynolds doing voice over work, but Fox was still disinterested. Two years later the footage was leaked, and the enormous positive reaction finally prompted Fox to put the movie into production. Despite this, Miller still wasn't the first choice for director; David S Goyer was initially approached. Negasonic Teenage Warhead was used for the film simply because the production team liked her name. Deadpool creator Rob Liefeld makes a cameo appearance in the film, and Fabian Nicieza is referenced by way of a prominent street sign. Daniel Cudmore was offered the role of Colossus, but turned it down as he wouldn't be voicing the character. Olivia Munn auditioned for Vanessa, but was offered Psylocke in **X-Men: Apocalpyse** instead. The opening credits don't give proper names, but rather descriptions of various characters, such as Ryan Reynolds' "God's Greatest Idiot" and Colossus' "CGI Character". There are a number of jokes at the expense of other Reynolds' projects, including a barbed comment about Green Lantern's costume, and an action figure based on Deadpool in his **Wolverine** appearance. Initially a number of other characters were to appear, including the villain Wyre, Wolverine and Taskmaster (budget cuts caused their removal, which led to Deadpool's joke about there being no other X-Men in the film because of the budget). Hugh Jackman does not appear in the film (the first "X" film in which that happens), but is mentioned several times. Deadpool – like the comic book version – makes many meta jokes, including asking Colossus which Professor he's being taken to; McAvoy or Stewart. The post credits sequence is a direct lift from **Ferris Bueller's Day Off**. The final battle takes place on a SHIELD Helicarrier, and Kevin Feige is referenced through pizza boxes from the company Feige's Favorites.

**Comic Notes:** *Vanessa Carlysle:* Although only credited as Vanessa, Morena Baccarin is playing Vanessa Carlysle, a mutant shapeshifter. Technically Vanessa first appeared in *The New Mutants #98 (Feb, '91)*, though she was actually pretending to be Domino at the time, which is who everyone thought she was. She appears properly – and using her usual codename Copycat – in *X-Force #19*. She was created, like Deadpool, by Fabian Nicieza and Rob Liefeld. Like the movie version she was a prostitute who fell in love with Deadpool, but when he left her because of the cancer she became a mercenary. Initially sent to kill Cable by a man named Tolliver, she infiltrated X-Force as Domino, but couldn't do it. She fell in love with Garrison Kane, but left him to return to Deadpool, then left him so he could pursue a relationship with Siryn. She teamed up with Sabretooth who later killed her, and she died in Deadpool's arms. What she didn't know was that he injected her with his blood, and so she healed and now runs a chimichanga stand.

*Ajax:* first appeared in *Deadpool #14 (Mar, '98)*, created by Joe Kelly and Walter McDaniel. His first name is really Francis, though his surname was never revealed. His powers are different to the movie version: in the comics he has enhanced strength and intuitive capacity, though thanks to implants he also has a high tolerance to pain, and superhuman agility. His backstory is not dissimilar to what we see in the movie; he worked for Weapon X and was the enforcer there, ultimately ordered to kill Wade Wilson. His torture and murder of Wade effectively created Deadpool, and the two began a bitter rivalry as they tried to kill each other. Unsurprisingly Deadpool succeeded.

*Jack Hammer:* or as he is better known Weasel, first appeared in *Cable #3 (Jul, '93)* created by Niceza and Joe Madureira. He is probably Deadpool's best friend, but he has a completely different back story to the movie – rather than a bartender, he's an arms dealer who works with several people, and who met Deadpool not long after he escaped the Weapon X program. The two shared an awkward friendship, with Deadpool abusing Weasel often, and Weasel equally taking advantage of Deadpool where he could. The two later fell out, and Weasel was employed to take down Deadpool, which he was almost able to do, but Deadpool escaped and later killed him.

*Blind Al:* was created by Kelly and Ed McGuinness for *Deadpool #1 (Jan, '97)*, where she was given a complex backstory that doesn't really appear in the movie. Originally a member of British Intelligence, Wade Wilson was hired to kill her but chose not to, and after becoming Deadpool he effectively kidnapped her, whereupon she became his housekeeper and flatmate. The pair have a very bizarre relationship – at times they hate each other and he is flat out abusive towards her, but at other times it seems they can't live without the other. Given the chance to go free, Al chose not to take it, though currently it would seem they don't live together.

*Negasonic Teenage Warhead*: Ellie Phimster first appeared in *New X-Men #115 (Aug, '01)* created by Grant Morrison and Frank Quitely. She is nothing like her movie version, despite a vaguely similar gothic appearance. Her powers, in the comics, are precognition, and she was essentially a background character who was later killed by the Sentinels. She was resurrected by Selene, using the Transmode Virus, but was destroyed when she refused to obey the Black Queen.

*Angel Dust*: is another background character who first appeared in *Morlocks #1 (Jun, '02)*, created by Geoff Johns and Shawn Martinbrough. Her powers are much the same as the movie version, but the comics' version ran away after her powers manifested, to join the Morlocks. When the Morlocks made a pact to help each other with one "aboveground issue", Angel Dust – or Christine – told her parents she was a mutant, but was surprised to find they were very accepting, and so she returned home. She later lost her powers when Scarlet Witch effectively removed all mutants from the Earth.

*And*: a character appears towards the end of the film – one of Ajax's henchmen – named Bob. This is a reference to Bob, Agent of Hydra (though of course, film rights mean Bob can't be an agent of Hydra!). First appearing in *Cable & Deadpool #38 (May, '07)*, and created by Niceza and Robert Brown, he has made quite a few appearances. He joined Hydra because his wife wanted him to have a proper job, but was tortured into helping Deadpool, and surprisingly became his sidekick. He went on a number of missions with Deadpool and various associates, and this impressed his wife no end. As time went on, Bob would be called upon to assist Deadpool if necessary, and he actually seemed to earn a place in Deadpool's heart, as the mercenary has saved his life on occasion.

**Ratings:** *IMDB:* 87%; *Rotten Tomatoes:* 84%; *Metacritic:* 65

**Review:** It's impossible to give this movie too much praise. From the opening titles through to the pastiche post credits sequence, this movie is a labour of love, giving Reynolds and Miller the chance to bring Deadpool to life perfectly. In fact, this is such a great representation of the character, it would have to be definitive. Reynolds' is the movie's greatest gift, but the rest of the cast nail their characters and Miller's direction is flawless. Nothing about this movie falters. A must see.

## X-MEN: APOCALYPSE
### *(27/5/2016)*

Thousands of years ago, a mutant named Apocalypse came into being, able to transfer his consciousness into other mutants and absorb their powers each time. Beaten during the building of the Pyramids, he is unearthed in 1983 by a cult that has built up around him, whereupon he recruits four mutants to serve him; one of whom is Magneto, whose family is killed when the locals discover his true identity. Mystique joins forces with Xavier as a new group of X-Men are formed in order to stop Apocalypse's plan to take control of the entire world.

**Cast:** *James McAvoy (Charles Xavier), Michael Fassbender (Erik Lehnsherr/ Magneto), Jennifer Lawrence (Raven /Mystique), Nicholas Hoult (Hank McCoy/Beast), Oscar Isaac (En Sabah Nur/Apocalypse), Rose Byrne (Moira MacTaggert), Evan Peters (Peter Maximoff/Quicksilver), Josh Helman (William Stryker), Sophie Turner (Jean Grey), Tye Sheridan (Scott Summers/Cyclops), Olivia Munn (Betsy Braddock/Psylocke), Lucas Till (Alex Summers/Havok), Kodi Smit-McPhee (Kurt Wagner/Nightcrawler), Ben Hardy (Angel), Alexandra Shipp (Ororo Monroe/Storm),* Zeljko Ivanek (Pentagon Scientist), *Tómas Lemarquis (Caliban), Lana Condor (Jubilation Lee/Jubilee)*

**Dir:** Bryan Singer; **Writers:** Simon Kinberg, Dan Harris, Michael Dougherty, Bryan Singer; **Prod:** Lauren Schuler Donner, Bryan Singer, Simon Kinberg; **Music:** John Ottman; **Exec.Prod:** Todd Hallowell, Josh McLaglen, Stan Lee; **DOP:** Newton Thomas Sigel; **Prod.Des.:** Grant Major; **Ed:** John Ottman ACE, Michael Louis Hill; **Costumes:** Louise Mingenbach; **20th Century Fox/ Donners' Company/Bad Hat Harry Productions/Marvel Entertainment/TSG Entertainment/Kinberg Genre**; 144; $178m; $544m

**Stan Spotting:** Playing Stan Lee watching nuclear weapons firing into the atmosphere. Rather sweetly, he's joined by his wife, Joannie this time.

**Should I Stay To The End?** Only if you're a comic fan. There's a scene which appears to set up the appearance of a villain to come.

**Notes:** Naturally, Hugh Jackman has a cameo as Weapon X (or Logan), complete with the actual Weapon X costume from the comics. John Ottman also appears as "confused tech". James Malloch's "Mystery Man" represents

Essex Corp in the post credits sequence, where he takes a sample of Weapons X's blood. In the comics, Nathaniel Essex is the supervillain Mr Sinister, who causes a lot of pain for the Summers' family. Sasha Pieterse was offered the role of Jean Grey and Taron Egerton was offered Scott Summers. Obviously both turned the roles down. This heavily rewrites the continuity of the X-Men movie series, apparently due to the altered timeline created in **Days of Future Past** (qv). As such a number of characters appear much earlier than they did in the original timeline. Quite what events from the original timeline still took place remain unclear.

**Comic Notes:** *Apocalypse:* The mutant who we would later discover was originally called En Sabah Nur, first appeared in *X-Factor #5 (Jun, '86)*, though like many this was just a shadowy cameo, and his full appearance was the following issue. Created by Louise Simonson and Jackson Guice, Apocalypse was, curiously, a late replacement for the Owl, who was intended to be the big bad that original X-Factor writer Bob Layton planned. The character in the movie is ambiguous in regards to his (original) abilities, but the comic version has telekinesis, telepathy, matter manipulation (specifically total control over his own physicality), super strength and speed, technopathy and regenerative abilities. The comic version looks a little like the movie version, but tends to be larger and bulkier. A villain with the aim of "survival of the fittest", the character was retconned so that he was involved in the Marvel universe since the time of the Pyramids, and influenced many events throughout history, including building alliances with the Hellfire Club and Mr Sinister. Since his appearance in the modern age, he has four bodyguards who he brands his Horsemen (again that's unclear in the movie – particularly as he appears to have them in **First Class** (qv)); four genetically modified mutants who are renamed War, Famine, Pestilence and Death. When Angel is given metal wings in the movie, this is a nod to the comics were a similar thing happened, and Warren Worthington took on the role of Death (the first version to be seen in the comics, though later it would be revealed that there were Horseman before this particular incarnation). Apocalypse has played a significant role in mutant comic storylines, notably *Fall of the Mutants* and *Age of Apocalypse* which played a huge role in the comic series in the mid-nineties.

    *Caliban:* was a member of the Morlock community (the underground of mutants who hide society) with an appearance quite similar to the movie version. He first appeared in *The Uncanny X-Men #148 (Aug, 81)* and would later become a member of X-Factor when Warren Worthington left after the Mutant Massacre. Although no power is suggested in the movie, in the comics Caliban can sense other mutants, and also, when terrified, can increase his size and strength and amplify the fear around him. Apocalypse would later take

Caliban and enhance him so he had access to those additional abilities all the time. It should come as no surprise to anyone that Caliban was killed and then briefly resurrected (although it was brief, and he is currently still dead).

**Ratings:** *IMDB:* 74%; *Rotten Tomatoes:* 48%; *Metacritic:* 52

**Review:** This movie is a little hard to review in that I'm quite invested in 80's X-Men (my favourite era) and Apocalypse (a brilliant villain). The movie doesn't come close to the brilliance of the comics in either department. Apocalypse has a muddled goal, which ultimately just turns out to be a desire to take over Xavier's body, while most of the mutants are nothing more than bodies with codenames; there is no characterisation for Angel and Psylocke at all. Sophie Turner and Tye Sheridan are badly miscast, while the decision to kill off Alex Summers seems a bizarre one. And sadly, this time, Quicksilver's scene seems very forced. In all, very disappointing.

## LOGAN
### *(3/3/2017)*

In 2029, Logan spends his days driving to make money to keep Professor Xavier and Caliban. Xavier's mental problems bring danger to all, but it is when Logan is contacted by a woman who has a child that Xavier has babbled about that brings them to the attention of the cyborg Donald Pierce. The girl is a clone of Wolverine and is the first of a new breed of mutants that must be protected. With no choice, Logan takes the girl and Xavier and goes on the run, Pierce and his Reavers close behind.

**Cast:** *Hugh Jackman (Logan/Wolverine), Patrick Stewart (Professor Charles Xavier), Richard E Grant (Dr Rice), Boyd Holbrook (Pierce), Stephen Merchant (Caliban) and introducing Dafne Keen (Laura)* , Elizabeth Rodriguez (Gabriela), Eriq LaSalle (Will Munson), Elise Neal (Kathryn Munson)

**Dir:** James Mangold; **Writers:** Michael Green, Scott Frank, James Mangold; **Prod:** Lauren Schuler Donner, Simon Kinberg, Hutch Parker; **Music:** Marco Beltrami; **Exec.Prod:** Stan Lee, James Mangold, Joe Caracciolo Jr, Josh McLaglen; **DOP:** John Mathieson; **Prod.Des.:** François Audouy; **Ed:** Michael McCusker, Dirk Westervelt; **Costumes:** Daniel Orlandi; **20th Century Fox/The Donners' Company/Marvel Entertainment/Kinberg Genre/TSG Entertainment**; 137; c $110m; $619m

**Notes:** Although initially claimed to be based on the "Old Man Logan" comic series, in truth only element are lifted from the comic – the idea of no heroes, Logan being old, going on a journey. Apart from that the story is vastly different. The X-Men comics that appear in the movie are not genuine comics, but are fakes, with art by Joe Quesada, requested by Jackman. Dan Panosian did the covers of the comics. In order to get Fox's approval for an R rating, Jackman was required to take a paycut. The casting of a new Caliban was due to the two film production teams working independently. The film that Xavier and Laura watch is **Shane**, a 1953 cowboy film.

**Comic Notes:** *Laura*: X-23 (as she simply is in the comics), came from a rather interesting creative background. She actually made her first appearance in the *X-Men: Evolution* cartoon, created by Craig Kyle. In terms of comics, she finally appeared in *NYX #3 (Feb, '04)*. Appearance and power-wise she is quite similar to the movie version, though in the comics she is older. Her origin story is roughly the same, though the program was headed by Martin Sutter, and Sarah Kinney was her creator, and was her surrogate mother. Zander Rice used radiation to force Laura's powers to manifest, and also created a trigger scent that drives her into a murderous rage. After killing Kinney at Rice's manipulation, she ended up on the streets, mostly mute and working in slum areas. She has been involved in a large number of storylines, including becoming Captain Universe, surviving the House of M Decimation, and leading an attack on Hulk. Currently, however, she has taken her father's mantle as Wolverine.

*Donald Pierce & The Reavers*: Donald Pierce, like his movie counterpart, is also a Cyborg and has led the Reavers. He first appeared in *Uncanny X-Men #132 (Apr, '80)*, and was created by Chris Claremont and John Byrne. Byrne stated that his appearance was based on Donald Sutherland, while Claremont revealed that his got his surname from the *M\*A\*S\*H* character Hawkeye Pierce. Pierce started as the White Bishop of the Hellfire Club, before becoming the White King, whereupon he revealed that he despised mutants – and indeed harboured much self-hate for his own cybernetics – and subsequently attempted to wipe out the Club. The X-Men stopped him and handed him over to the Club. He escaped the Club and resurfaced as the leader of the Reavers – a trio of cyborgs he created – and joined forces with Lady Deathstrike and three mercenaries – Cole, Macon and Reese who he also enhanced – to wage war on the Hellfire Club and the X-Men. They were beaten, and Pierce was left as little more than a head. He popped up from time to time to cause trouble, and when becoming a prisoner of the Young X-Men formed a surprising bond with Dust. He was ultimately killed by Cyclops.

*Zander Rice*: first appeared in *X-23 #1 (Mar, '05)*, and was created by Kyle, Christopher Yost and Billy Tan. Surprisingly he is very similar to his movie version, even down to Logan having killed his father. In the comics, though, he is crueller, and treated Laura quite badly, calling her an animal and having her kill Sarah Kinney.

**Ratings:** *IMDB:* 82%; *Rotten Tomatoes:* 93%; *Metacritic:* 77

**Review:** Raved about as one of the greatest comic book movies of all time, it's hard to deny the brilliance of the film. In truth there's something tremendously clever in the way that Logan and Xavier deal with their powers in old age, and the character of Laura is brilliant, meshing fantastically with Logan. Jackman does, however, steal the entire movie, though the rest of the cast is universally brilliant. It's an oddly meta-film that does little to bring coherence to Fox's X-universe, but as a stand-alone film, it's singularly spectacular, and worthy of all the praise.

## Is the Return of the X-Men to Marvel a Good Thing?

Sony's decision to work with Marvel on a Spider-Man movie (and the subsequent huge success of that deal) immediately resparked the wishes of many vocal advocates that the X-Men and Fantastic Four should be returned to Marvel and incorporated into the MCU. When Disney revealed that they were in the process of acquiring Fox's television and movie arms, it seemed that fans would get their wishes, and yet it's hard to know if this is going to be a good thing or not.

The idea of the Fantastic Four returning to Marvel and entering the Marvel Cinematic Universe is a tempting one, as Marvel Studios can't make a worse FF film than Fox have, and the team would fit in rather nicely there. The X-Men, on the other hand, one would have to argue perhaps not. Whilst Fox's X-movies are a bit hit and miss, Fox's freedom from Disney's happier MCU does allow them to make films like **Deadpool** and **Logan**; if the X-franchise were in Marvel's hands, those films simply wouldn't have happened. In order for Marvel to continue to produce output as Fox is doing, they would have to push the X-series into television only, and mostly on Netflix where the company can go as dark as they like (and, let's face it, while the idea of several X-Men television series isn't a particularly bad one, Disney's decision to withdraw its properties from Netflix now render that moot. It's not the end, though, as Hulu's plans for **Helstrom** suggest there's still potential for Disney to go dark).

But if **Inhumans** has proved anything, it's that team-ups of super powered individuals really do need a big budget behind them to show off.

*Agents of SHIELD* have used Ghost Rider well, but the cost of the effect relegated him to a third of the season he appeared in. *The Gifted* has given us a group with powers, but those powers are mostly practical effects, rather than astonishing light and sound shows. In order to look as good as they should, the X-Men need to be in films. And Fox has more flexibility with its tone in film than Marvel does.

But that's not the only reason X-Men staying with Fox was generally a good idea. There's a limit to Marvel Studios output, and at the moment that is three films a year. If the X-Men were to return to Marvel, we'd have to wait some time before we'd get another X-film, and then it would be simply an X-Men film, possibly once every two-three years. Fox had the ability to churn out so much more, and while Marvel might be looking for other properties after **Avengers: Endgame**, the reduced output and the cutting of very dark movies would be enough to support the idea that the X-Men may not thrive at Disney.

## DEADPOOL 2
### *(1/6/2018)*

Deadpool's missions bring him into conflict with a villain who has Vanessa murdered, driving him to the pits of despair and forcing him to join the X-Men. However, when he encounters a boy named Russell who is being abused at an orphanage, Deadpool's response results in the pair being incarcerated with power inhibitors. As Deadpool faces death without his regenerative abilities, a mutant from the future – Cable – arrives to kill the mutant who will murder his family.

**Cast:** *Ryan Reynolds (Wade Wilson/Deadpool), Josh Brolin (Cable), Morena Baccarin (Vanessa), Julian Dennison (Firefist), Zazie Beetz (Domino), T J Miller (Weasel), Leslie Uggams (Blind Al),* Karan Soni (Dopinder), *Brianna Hildebrand (Negasonic Teenage Warhead), Jack Kesy (Black Tom),* Eddie Marsan (Headmaster), *Shioli Kutsuna (Yukio), Stefan Kapicic (Colossus' Voice),* Randal Reeder (Buck), Nikolai Witschi (Head Orderly Frye), Thayr Harris (Sergei Valishniko), Rob Delaney (Peter), *Lewis Tan (Shatterstar), Bill Skarsgard (Zeitgeist), Terry Crews (Bedlam)*

**Dir:** David Leitch; **Writers:** Rhett Reese, Paul Wernick, Ryan Reynolds; **Prod:** Lauren Schuler Donner, Simon Kinberg, Ryan Reynolds; **Music:** Tyler Bates; **Exec.Prod:** Jonathon Komack Martin, Kelly McCormick, Rhett Reese, Ethan Smith, Aditya Sood, Paul Wernick, Stan Lee; **DOP:** Jonathan Sela; **Prod.Des.:** David Scheunemann; **Ed:** Craig Alpert, Michael McCusker, Elísabet Ronaldsdóttir, Dirk Westervelt; **Costumes:** Kurt Swanson, Bart Mueller; **20th**

**Century Fox/TSG Entertainment/Maximum Effort/The Donners' Company/ Kinberg Genre/Marvel Entertainment**; 119; $110m; $786m

**Stan Spotting:** Just a picture this time – he is on a billboard as Domino parachutes in.

**Notes:** Brad Pitt and Matt Damon have uncredited cameos in the films, the former as the Vanisher, the latter as a heavily made up redneck. There are a plethora of easter eggs in the film, including the fact that Deadpool's swords are named "Bea" and "Arthur" (inscribed on the bases); the actors playing X-Force filmed action sequences for the trailer to hide their ultimate fates in the movie; Cable's gun going to "11" is an obvious reference to **THIS IS SPINAL TAP**, though the poster "Puppet Show At Noon" is another less obvious one; the title sequence is a deliberate nod to the James Bond films; there are a number of references to the first film (unsurprisingly), but the best is possibly Deadpool lifting a floorboard in Blind Al's house to reveal cocaine and the cure for blindness. Brad Pitt's pay for the film was a cup of coffee. On December 12th, the film was rereleased as **ONCE UPON A DEADPOOL**, in which heavy editing took place to make the film more family friendly, and Fred Savage joined the cast, being told the story of the film by Deadpool (in a homage to **THE PRINCESS' BRIDE**). The film contained some additional material (though much has already been seen on the extended version released on home entertainment), and some of the profits were dedicated to charity. Initially Black Tom Cassidy was to have been the villain of the movie, but it was later decided that a bigger threat was needed, and so Juggernaut was used. Reynolds voiced the part, while the face was digitally modelled on Leitch.

**Comic Notes:** *Cable:* Oh boy. If there was ever a character that epitomised the complex and convoluted backstories of the X-Men, then Cable is absolutely that character. In physical appearance, the comic version is very, very similar to the movie version. Personality, they are similar to, but in terms of backstory, they are radically different. Cable is Nathan Christopher Charles Summers, the biological son of Scott Summers (Cyclops) and Madelyne Pryor (the clone of Jean Grey created by Mr Sinister); he is part of the Summers dynasty. Originally Cable was just a time traveller from the future, fighting his archenemy Stryfe (who would turn out to be a clone of Cable), but it was later revealed that Nathan – as a baby – was infected by the techno-organic virus, and was transported to the future so he could be cured (as an aside, Mother Askani transported him through time, and she was actually Rachel Summers – the daughter of Scott Summers and Jean Grey; and Askani would then drag Scott and Jean's minds to the future to raise Nathan as Slym and Redd during which time they would discover that Sinister had prompted the creation of Nathan to defeat Apocalypse. Told you it was complex...). Cable would then return to the present day and turn the New Mutants into X-Force, before teaming up with Deadpool, but would ultimately succumb to the techno-organic virus and die, saving X-Force. Being the X-Men, Cable returns several more times, before being killed again. This time by his younger self. No doubt he'll be back. Nathan Summers first appeared in *The Uncanny X-Men #201 (Jan, '86)* created by Chris Claremont, but the Cable version – created by Louise Simonson and Rob Liefeld – made his debut in *The New Mutants #87 (Mar, '90)*.

*Domino:* Domino technically first appears in *X-Force #8 (Mar, '92)* (though it's more a cameo; she doesn't fully appear until #11), though Copycat pretends to be her much earlier in *The New Mutants #98 (Feb, '91)*. She was created by Fabian Nicieza and Rob Liefeld, and unlike the movie version, is Caucasian, with a black circle around her left eye (obviously swapped for the movie). In the comics she is quite closely connected to Cable, being romantically connected to him. Her real name was Neena Thurman, and she was created to be the perfect weapon, though her luck power was deemed a failure, and her biological mother managed to free her from the company that created her. She met Cable, created the Wild Pack, joined X-Force, dated Risque, joined the X-Corporation, fought Cable and Deadpool, re-joined the Wild Pack (or Six Pack), dated Warpath...it's a complex history. She is currently working with Sabretooth as part of Weapon X-Force.

*Firefist:* Russell Collins – better known as Rusty – is quite different to his movie counterpart. The comic version is a late teens, white guy with red hair, and a much more impressive physique. He first appeared in *X-Factor #1 (Feb, '86)*, created by Bob Layton and Jackson Guice, and has a fairly interesting story – handed over to X-Factor after killing a woman accidentally when his powers manifested, Rusty became close friends with the other X-Factor wards – particularly Skids – and together they formed the X-Terminators, before joining the New Mutants. He and Skids were later brainwashed, before being rescued and restored by Magneto, but he ultimately died when the mutant Holocaust drained his life essence.

*Black Tom:* Tom Cassidy (created by Chris Claremont and Dave Cockrum in *X-Men #99 (Jun, '76)*) is the cousin of Sean Cassidy, or Banshee. The movie version, though physically similar to the comic version, doesn't give us any indication of his powers, but in the comics he has the ability to fire concussive blasts. He is a frequent opponent of Deadpool and ally of Juggernaut, but turned to crime when his wife Maeve was killed in an IRA bombing (something he blamed Sean for). Later he would become infected with a wood-like substance that virtually turned him into a tree, but this was reversed. Tom, having gone virtually insane, would later kill a young boy who had helped turned Juggernaut to good. When Tom lost his tree powers, Juggernaut convinced him to turn himself into the police for the murder of the boy.

*Shatterstar, Zeitgeist & Bedlam*: Gaveedra-Seven, Larry Ekler and Jesse Aronson, in truth, bear very little resemblance to their comic book counterparts, though Shatterstar does match physically. Created by Fabian Nicieza and Rob Liefeld in *the New Mutants #99 (Mar, '91)*, Shatterstar can create shockwaves, but was originally from Mojoworld (a universe created by the cross-dimensional being, Mojo). He did join X-Force when Cable created it, and would go on missions with them, but his past remained a murky area, though it was eventually revealed he was the son of X-Men Dazzler and Longshot. He seemed to briefly have a relationship with a mutant named Rictor, but this was complicated by his desire to explore his newfound passion. Ekler originally went by the name Everyman, and is a swordsman, but was more of a villain, killing superheroes while under the influence of Doctor Faustus. He first appeared in *Captain America #267 (Mar, '82)*, created by JM DeMatteis and Michael Zeck, and later died at the hands of Vormund, impaled on his own sword. Bedlam first appeared in *Factor X #1 (Mar, '95)*, created by John Francis Moore and Steve Epting, and has the ability to generate an electromagnetic field which disables electronics. He joined X-Force after meeting Domino while trying to track down his brother. This he was successful at, but sadly he found his brother had killed his parents with his own mutant ability, and needed to be stopped. Though he survived the infamous M-Day (where most mutants lost their abilities), it's currently unknown where Bedlam is.

*The Vanisher*: is not an invisible mutant hero that looks like Brad Pitt when he is electrocuted. In fact, the comic book version couldn't be more different. He's actually a villain who first appeared way back in *The X-Men #2 (Nov, '63)* created by Stan Lee and Jack Kirby. He actually has the power to teleport and has been an opponent of the X-Men and it's various sub-groups for over fifty years, but reluctantly served with X-Force, often getting shot as a result (including once by Deadpool). He recently turned back to crime (possibly because of brainwashing) and was arrested for smuggling Vibranium out of Wakanda.

*And:* Juggernaut appeared in **X-Men: The Last Stand**, if you want his backstory, while Yukio may or may not be the same character who appeared in **The Wolverine**; certainly she's nothing like the comic character.

**Ratings:** *IMDB:* 78%; *Rotten Tomatoes:* 83%; *Metacritic:* 66

**Review:** There is a certain sameness to this film, which is inevitable insofar as it's impossible to recapture the originality of the first film. However, the jokes come thick and fast, and the swagger of the first film is replaced by a confidence in just what they film can do. The decision to introduce and then kill off X-Force is tremendously funny, and the cameos are downright hilarious. Sadly the female characters don't get as much screen time as they deserve (the film could do with a lot more Domino, and Morena Baccarin's absence for much of the film is notable), but Josh Brolin's performance compliments Reynolds' perfectly.

## VENOM
### *(5/10/2018)*

Trapped in a self-destructive life, Eddie Brock becomes obsessed with Carlton Drake and the projects he is exploring, little knowing that Drake has acquired a number of alien life forms that can symbiotically bond with human beings. When one bonds with Brock, he discovers a creature called Venom who wants to absorb life on Earth. But Venom has a boss – Riot – and in order to stop him, Eddie realises he has to find a way to work with Venom.

**Cast:** *Tom Hardy (Eddie Brock/Venom), Michelle Williams (Anne Weying), Riz Ahmed (Carlton Drake/Riot), Scott Haze (Roland Treece),* Reid Scott (Dr Dan Lewis), Jenny Slate (Dr Dora Skirth), Melora Walters (Homeless Woman Maria), *Woody Harrelson (Cletus Kassady)*

**Dir:** Ruben Fleischer; **Writers:** Jeff Pinkner, Scott Rosenberg, Kelly Marcel; **Prod:** Avi Arad, Matt Tolmach, Amy Pascal; **Music:** Ludwig Goransson, **Exec.Prod:** Howard Chen, Edward Cheng, Tom Hardy, David Householter, Kelly Marcel, Stan Lee; **DOP:** Matthew Libatique; **Prod.Des.:** Oliver Scholl; **Ed:** Alan Baumgarten, Maryann Brandon; **Costumes:** Kelli Jones; **Colombia Pictures/Tencent Pictures/Arad Productions/Matt Tolmach Productions/ Pascal Pictures/Marvel Entertainment**; 112; $110m; $856m

**Stan Spotting:** A dog walker at the end of the film who tells Eddie he'll work it out.

**Notes: Venom** has been in the works for a considerable time, with the intention to make it a spin-off film from **Spider-Man 3**, though this was abandoned when the character (and the movie) failed to live up to expectations. Talk of the film continued for quite some time, and resurfaced when **The Amazing Spider-Man** was made, but the decision to hand Spider-Man over to Marvel again scuppered this idea. With **Spider-Man: Homecoming** looking as good as it did, Pascal and Sony again revived the idea of a Venom film. Amy Pascal initially claimed that **Venom** was part of the MCU, but this was quickly denied by Kevin Feige. Anne works for Michelinie and McFarlane, a reference to the creators of Venom. Tom Holland did film a cameo for the film as Peter Parker, but it was removed at the request of Kevin Feige, who acted as an unofficial advisor to the film.

**Comic Notes:** *Anne Weying*: first appeared in *The Amazing Spider-Man #375 (Mar, '93)*, created by David Michelinie and Mark Bagley, and is indeed a lawyer and ex of Eddie Brock (though wife in this case). She helped Spider-Man in his battle against Venom, and was able to talk her ex-husband down. Notably, however, she was once shot, and the Venom symbiote bonded with her to save her life, whereupon she became She-Venom for a short while (something the film gives a little nod to). She gave the symbiote back to Eddie when he needed it to save his life, but realised that Eddie was utterly dependent on it. Rather sadly, on seeing Spider-Man in the black costume and still recovering from being She-Venom, she lost touch with reality and killed herself.

*Carlton Drake*: is a good choice for villain, as he tends to be an opponent of both Spider-Man and Venom. He is the head of the Life Foundation, and was responsible for creating a number of other Symbiotes, including Riot, though he never bonded with the symbiote. He did become a sort of Man-Spider creature after injecting himself with Spider-Man's blood, though he has since recovered, but disappeared. He first appeared in *The Amazing Spider-Man #298 (Mar, '88)*, created by Michelinie and Todd McFarlane.

*Roland Treece*: is different in the comics to his movie counterpart – he was the CEO of Treece International, though he did work with Carlton Drake as he was on the board of the Life Foundation. He was once stopped by Venom while trying to recover a hoard of gold, and helped turn Carlton Drake into Homo Arachnis. He first appeared in *Venom: Lethal Protector #3 (Apr, '93)*, created by Michelinie and Bagley.

*Cletus Kasady*: Kasady first appeared in *The Amazing Spider-Man #344 (Mar,'91)* created by Michelinie and Erik Larsen. He is a horrible human being, born in the prison Ravencroft, and going on to murder a large number of people as a child. His abuse at St Estes Home for Boys (where he was sent after attempting to kill his mother) led him to believe that there was no meaning to life, except killing people. When sharing a cell with Eddie Brock, Venom's offspring bonded to Kasady and he became Carnage (you might be able to see where the sequel will be going at this point). As Carnage, Kasady has confronted many superheros, notably Spider-Man, Deadpool and Venom, but has always been stopped. Nonetheless, the character has become as popular as Venom, even getting his own series.

*Riot:* was, as mentioned above, created by Carlton Drake from Venom, along with Scream, Phage, Lasher and Agony, and bonded with Trevor Cole, before being stopped by Spider-Man and Venom. He has appeared since then, bonding with his other symbiotes, as well as Deadpool, but is inevitably defeated each time. He made his first appearance in *Venom: Lethal Protector #4 (May, '93)*, created by Michelinie and Ron Lim.

*And:* Venom appeared in **Spider-Man 3**, where his comic notes can be found, though it should be said that the Tom Hardy version of the character is closer to the comic version than the Topher Grace version. The movie itself is based very loosely on (as you may have guessed from the above notes) *Venom: Lethal Protector*, and also "Planet of the Symbiotes", a 1995 *The Amazing Spider-Man* storyline.

**Ratings:** *IMDB:* 68%; *Rotten Tomatoes:* 28%; *Metacritic:* 35

**Review: Venom** suffers for the same reason that **Aliens Vs Predator** and **Freddy Vs Jason** suffers – ultimately a villain becomes good, which, at the end of the day, it's not really something the audience wants to see. The problem is, this film could work well, if it spent more time concentrating on Eddie Brock trying to tame Venom, and ultimately making a deal with him in order to save Earth; in essence the Jekyll versus Hyde aspect that Hardy hyped up. Equally the film deserved to be an R-rated horror (which seems to be what the opening suggests). Without these, the film lacks bite and when Venom ultimately reveals he's helping Eddie because they're both losers, it's far too easy to check out. A missed opportunity, surprisingly.

## Is Sony's Spider-verse Working?

Having come to an uneasy alliance with Disney, Sony seems to remain committed to having Marvel making Spider-Man movies, of which Disney and Sony share the profits. In truth, and with historical evidence, this is a good idea for both sides of the fence. The MCU without Spider-Man, nowadays, seems a little weird. Equally, Sony hasn't really done great things with the character after a couple of brilliant films. Now, it seems, we have the best of both worlds.

However, at the same time, Sony are slightly disadvantaged by this particular deal. Marvel can churn out a Spider-Man film every two years, but while they aren't doing that, they are popping out other Marvel films which make them a lot of money. After **No Way Home**, there's quite the break with

Spider-Man, and this doesn't really affect Marvel in any significant way. If they aren't making new films, they have a slew of television series to earn them cash.

Sony, however, don't really have much in their store at the moment, and they need a good franchise to keep them in the green. Creating a Spider-verse (albeit one that doesn't actually feature Spider-Man) seemed to be a good idea. However, Sony seem to be fumbling with this particular franchise in a way that they did with the Spider-Man movies of old.

After **Homecoming**, Sony announced a number of different projects, of which only one (**Venom**) actually materialised. Black Cat and Silver Sable (actually not bad ideas for films) remained talked about for a while, before they were suddenly not talked about anymore. **Venom**, of course, made a nice sum of money for Sony, but the critical reaction wasn't really positive. **Venom: Let There Be Carnage** suffered from an even more critical mauling and a significant drop in income. A third is on the way (allegedly), but the signs aren't positive. And the attempts to tie the Spider-verse into the MCU seemed very tacked on.

Until, **Morbius**, of course, where there was a very strong link, which is largely ignored, because **Morbius** did spectacularly badly at the box office.

Sony have got three more project ready and waiting, in the form of **Madame Web, Kraven The Hunter,** and the third **Venom** movie. All three films are slated for release in 2024, and as such, this will probably be the make-or-break year for Sony.

What's, perhaps, the most baffling thing about this situation is that Sony seems to be heading down the path of making a series of anti-hero films, rather than doing the obvious and making a superhero movie. It's clear that Sony can use Spider-Man in a film if they so choose, and at present they have opted not to (which is understandable – it's probably done to make peace with Marvel). But Sony have two other Spider-Men they could play with: Miles Morales and Gwen Stacey. Marvel may be working towards using Miles Morales, but it's baffling as to why Sony don't just bother to make a Spider-Gwen movie, and capitalise on the positivity that the animated Spider-Verse films have created.

If the history is any sign, it seems that the Sony-verse will come to a grinding halt next year, and more importantly, that Sony haven't learnt anything from their Spider-Man films and working with Marvel. And truthfully, given what Sony have to play with, this is a huge disappointment.

# DARK PHOENIX

## (6/6/2019)

A mission in space creates a situation where Jean comes into contact with an entity that she absorbs and Hank is shocked when he discovers her powers have become enhanced. When Xavier attempts to reach her mind, she rejects him, and goes to Magneto, attacking her father and killing Mystique in the process. However, it is her contact with Vuk, leader of the alien D'Bari that gives her the truth of her new found power, and its destructive capabilities.

**Cast:** *James McAvoy (Professor Charles Xavier), Michael Fassbender (Erik Lehnsherr/Magneto), Jennifer Lawrence (Raven Darkholme/Mystique), Nicholas Hoult (Hank McCoy/Beast), Sophie Turner (Jean Grey/Phoenix), Tye Sheridan (Scott Summers/Cyclops), Alexandra Shipp (Ororo Munroe/Storm), Kodi Smit-McPhee (Kurt Wagner/Nightcrawler), Evan Peters (Peter Maximoff/ Quicksilver) and Jessica Chastain (Vuk), Scott Shepherd (John Grey)*, Ato Essandoh (Jones), Brian d'Arcy James (President), *Kota Eberhardt (Selene)*, Andrew Stehlin (Ariki), *Halston Sage (Dazzler)*

**Dir/Writer:** Simon Kinberg; **Prod:** Lauren Schuler Donner, Simon Kinberg, Hutch Parker, Todd Hallowell; **Music:** Hans Zimmer; **Exec.Prod:** Josh McLaglen, Stan Lee; **DOP:** Mauro Fiore; **Prod.Des.:** Claude Paré; **Ed:** Lee Smith; **Costumes:** Daniel Orlandi; **20th Century Fox/TSG Entertainment/ Bad Hat Harry Productions/The Donners' Company/Kinberg Genre/Marvel Entertainment**; 114; $200m; $254m

**Notes:** This is the first film in the X franchise that makes no mention of Wolverine at all. Jennifer Lawrence agreed to appear in the film as long as Mystique was killed off. Lamar Johnson plays the mutant Match, in a brief appearance. Although unnamed, Magneto's mutant hideaway would appear to be Genosha (an actor is credited as Genosha Sentry) – in the comics, Genosha is an island that Magneto takes control of to provide sanctuary to mutants. Andrew Stehlin's character was originally Red Lotus, but was later changed. The initial draft of the story featured the return of the Hellfire Club and the Shiar Empire, as the story is based on the *Phoenix Saga* in the *X-Men* comics. A number of reshoots were undertaken to change the third act to avoid comparisons to **Captain Marvel**, as the climax would have occurred in space. With all these issues, the original release date of 2 November, 2018 was pushed back to 14 February, 2019. Allegedly, the response by China to the trailer saw it pushed back to being a summer blockbuster as at one point it was going to be a straight-to-DVD release. Strangely, Jean gets the Phoenix force in this movie, but appears to display it in **X-Men: Apocalypse**. Jean's opening narration is that of Xavier's in **X2**, while the shot of Blackbird coming up through the basketball court mimics a very similar shot in **X-Men**. Hans Zimmer had retired from movie composition, but was convinced by Kinberg to return. Internationally, the film was retitled **X-Men: Dark Phoenix**, presumably to make the connection to the other films clearer.

**Comic Notes:** *Vuk:* first appeared in *Avengers #4 (Mar, '64)*, created by Stan Lee and Jack Kirby, and usually used the name Starhammer. He (yes, he) crashed into Earth and later sought the help of the Avengers and Namor to escape, which he did. Later comics retconned him as a D'Bari, a species whose home world were destroyed by the Phoenix force. Though superhumanly strong, neither Vuk, nor the D'Bari in general were able to change form.

·*Selene:* is one of the oldest mutants on the planet – possibly the oldest. She is a psychic vampire, able to feed off the psyche of others to extend her life, which has granted her virtual immortality. She also possess telekinesis and telepathy, which makes her a little like her movie counterpart. She was invited to join the Hellfire Club, becoming their Black Queen, and became a constant thorn in the side of the X-Men and their allies, but she ultimately abandoned the Club but would return to them and take her place as Black Queen, though when Sebastian Shaw returned to the Club he locked her away for her earlier betrayal. She has frequently attempted to take control of Rachel Summers – the future, alternate daughter of Jean Grey and Scott Summers – but rarely succeeds. Most recently she has been freed from her imprisonment, was killed by Warpath, resurrected by Lady Deathstrike and Enchantress, and now works for the White House. She looks nothing like the movie version; rather she has a more traditional vampire look – pale, with long, dark hair. She first appeared in *New Mutants #9 (Nov, '83)*, created by Chris Claremont and Sal Buscema.

*Dazzler:* Alison Blaire looks almost identical to her movie counterpart, and has the same powers (the ability to convert sound into light), *and* is a singer. She's so close to the comic version, it's amazing she's barely in the film. In the comics she has a great backstory – a singer who used her mutant powers to create amazing light shows when she performed, though she hid her mutant nature, and turned down the X-Men's first offer to join them. She did frequently fight to help those in need, and later joined the X-Men after she was possessed by Malice. She fell in love with a fellow X-Man; Longshot (an extra-dimensional being), and then moved between the X-Teams, before joining SHIELD at Maria Hill's behest. Dazzler has since worked with A-Force – an all-female Avengers team - but found herself at odds with Carol Danvers when Nico Minoru was revealed to be about to kill someone, and Danvers wanted to arrest her. Most recently she supported Emma Frost against the

Inhumans. Dazzler first appeared in *The Uncanny X-Men #130 (Feb, '80)* created by Tom DeFalco and John Romita Jr. The reason for her creation, however, is rather fascinating. Casablanca Records were hoping to manufacture a new artist on a cross-media platform, starting with an animated film featuring "Disco Queen". Marvel were approached to create the character, while Casablanca would find the singer, and Editor-in-chief Jim Shooter wrote the script for the animation, which was then changed to be a live action film. DeFalco wrote her backstory and Romita designed her, while Roger Stern suggested changing the name to Dazzler. By now Casablanca Records had pulled out of the concept, but Filmworks were keen for it to go ahead, and wanted Bo Derek as Dazzler, but they could only get her if John Derek directed the film, and unhappy with this, Filmwork gave up, leaving Marvel to keep the character for the comics. For some time it was speculated that because of this, Fox didn't actually have the rights to use the character, but this was clearly not the case.

*Match*: Ben Hammil joined Xavier's School and under Rahne Sinclair, was the leader of the Paragon Squad. He has remained a background figure since "M-Day" (the event which cost most mutants their abilities) and was most recently seen attending the Jean Grey School of Higher Learning. As his head is permanently on fire in the comics, he is not tremendously like his movie counterpart. Match first appeared in *New Mutants Vol. 2 #7 (Jan, '04)*, created by Nunzio DeFillippis and Christina Weir.

**Ratings:** *IMDB:* 60%; *Rotten Tomatoes:* 22%; *Metacritic:* 43

**Review:** There is so much to dislike about this film, it's difficult to know where to start. Sheridan and Turner have not improved at all, and because we barely know these iterations of the characters, it's difficult to generate any love for them. Lawrence looks as though she wants to be anywhere but on set, while Peters is again written out quickly so no one has to query as to why his powers aren't always being used. Chastain is wasted as a villain that has no interesting qualities and very basic motivations. It's as though Kinberg hadn't learnt anything from the last disastrous time he wrote a Phoenix story. A terrible ending to the X-saga.

## NEW MUTANTS
### *(5/10/2018)*

When Dani Moonstar finds herself alone after her father is killed by a giant bear, she meets Cecilia Reyes who takes her in at a special hospital for those like her. Yet there is definitely something wrong in the hospital – her fellow patients believe that their super abilities require them to be trained to join the X-Men. When they realise they can't leave the hospital and are actually prisoners, and that they are being haunted by terrifying visions of their pasts, the five work together to find a way to escape the creature that is pursuing them.

**Cast:** *Maisie Williams (Rahne Sinclair/Wolfsbane), Anya Taylor-Joy (Illyana Rasputin/Magik), Charlie Heaton (Sam Guthrie/Cannonball), Alice Braga (Cecilia Reyes), Blu Hunt (Danielle Moonstar/Mirage), Henry Zaga (Roberto da Costa/Sunspot),* Adam Beach (William Lonestar)

**Dir:** Josh Boone; **Writers:** Josh Boone, Knate Lee; **Prod:** Lauren Schuler Donner, Simon Kinberg, Karen Rosenfelt; **Music:** Mark Snow; **DOP:** Peter Deming; **Prod.Des.:** Molly Hughes; **Ed:** Robb Sullivan; **Costumes:** Leesa Evans; **20th Century Fox/TSG Entertainment/Sunswept Entertainment/ Kinberg Genre/Marvel Entertainment**; 94; c $70m; $49m

**Notes:** Marilyn Manson provides the voices of the Smiley Men. Early drafts of the script saw it set three years after **X-Men Apocalypse** and featured Professor Xavier and Warlock. James McAvoy and Alexandra Shipp were attached to appear (a draft of the script featured Storm), but both characters were later dropped. Rosario Dawson was cast as Cecelia Reyes, but pulled out before production began. Henry Zaga is Brazilian, but not Afro-Brazilian like the character of Roberto da Costa, which caused some controversy. Josh Boone originally wanted the film to be a full on horror film, but this was toned down by various writers (the shooting script was co-written by Scott Frank, Josh Zetumer, Chad and Carey Hayes, Seth Grahame-Smith, Scott Neustadter and Michael H Weber). The film was shot in 2017, and the negative response to **Apocalypse** saw a number of changes, including moving the film to modern day and dropping Warlock. The plan was to make the film more Young Adult, but the success of **It** prompted the studio to go back to the horror version. The 2018 release date was changed to February 2019, and this was then pushed back to August, 2019. At this point the intention was to reshoot almost half the film. Jon Hamm was in talks to play Mr Sinister in a post credits sequence, though this was dropped, and Antonio Banderas was approached to play Emmanuel da Costa, Roberto's father. When Disney acquired Fox, the release date was pushed back to April, 2020. COVID forced the date to change once again to August, 2020. It would appear that no reshoots took place, and Banderas' never filmed his scenes either, as Boone suspected that there was no chance there would ever be a sequel. Marilyn Manson recorded a cover version of *Cry, Little Sister* for the film, but this was ultimately not used.

**Comic Notes:** *The New Mutants:* first debuted in *Marvel Graphic Novel #4: The New Mutants (Sep, '82),* created by Chris Claremont and Bob McLeod. The original team consisted of Cannonball and Mirage (the leaders), Sunspot, Wolfsbane and Karma. They were later joined by Cypher, Magik, Magma and Warlock. Originally they were students at Xavier's School for Gifted Youngsters, but when Xavier disappeared and Magneto took over as the new principal, the New Mutants started to rebel more and more. They would wear yellow and black uniforms, but had alternative superhero costumes when they broke the rules and went out on their own missions. They also had rivals in the Hellions, a group of mutants who were being taught by Emma Frost at her own mutant academy. Later they would be joined by Rusty, Skids, Boom-Boom and Rictor, who were mutants protected by X-Factor. (At the time there was a big division between the X-Men and X-Factor, and then Xavier disappeared, and the X-Men seemed to die, but actually went to Australia, and then New York fell into Limbo, and scientists got eaten by an elevator in a Daredevil comic...it was a whole thing). By the time Cable took over, the New Mutants had become a different group; this movie really concentrates on the original version. This original lineup has since reformed on occasion, with some of the more deceased members (Cypher...) being brought back to life.

*Rahne Sinclair:* was the daughter of a sex worker and a priest, in the comics, and was looked after by Moira McTaggert after her father discovered her powers (which are essentially what is seen in the movie) and led a lynch mob to kill her. She encountered Dani, and helped her rescue Roberto, before they were joined by Sam as they tried to fight Donald Pierce. This led to her joining the New Mutants using the codename Wolfsbane. She struggled with reconciling her faith and her powers, and in the comics was good friends with Dani and had a crush on Sam (which is slightly different to the movie...). After mentally bonding with Havok and being manipulated by the Shadow King, she joined X-Factor, often remaining in a semi-wolf form. Later she joined Excalibur, which helped her mature and come to terms with herself. She would later have a child, Tier, who Jack Russell adopted, though when Tier later dies she became a deaconess. She joined Magik's New Mutants, and was later beaten to death by an angry mob who hate mutants. Naturally she was resurrected.

*Illyana Nikolaievna Rasputina:* is not like the other characters in that she first appeared in *Giant Size X-Men #1 (May, '75),* the younger sister of Piotr Rasputin, the X-Man known as Colossus (though she is not named for another six years). She was originally created by Len Wein and Dave Cockrum, but Chris Claremont and Sal Buscema decided to change the character, and had her sent to Limbo where she remained for seven years (though time passes quicker in Limbo), and returned a teenage sorceress who had the mutant ability to create teleportation discs, and who had managed to become the queen of Limbo (all of which is revealed in a comic series called *Storm and Illyana*). She then appeared in the comic series *Magik* (where she took on her codename), and subsequently joined the New Mutants. She also gained a weapon called the SoulSword, and the more she used this the more armour appeared on her and the more demonic she became. After the demons S'ym and N'astirh were defeated in their attempt to take over Earth, Illyana lost the influence of Limbo, and regressed to being a little girl, with no mutant ability. She then contracted the legacy virus and died. Naturally she was resurrected – by Belasco, the man who taught her sorcery the first time, and as the Darkchilde she returned, though the Darkchilde persona often caused problems for the mutants. Though imprisoned for her behaviour she was later released and on Karma's instructions, reformed the New Mutants.

*Sam Guthrie*: discovered his powers when he was in a coal mine and they manifested, saving his life, and the life of his father's friend. This attracted the attention of Donald Pierce, who recruited him to the Hellfire Club, but he turned on Pierce to help Roberto da Costa, who he would later become firm friends with. He then became part of Xavier's School. He and Danielle Moonstar acted as the leaders of the New Mutants, and his honesty and integrity saw him join several different X- teams over the years. He also joined the US Avengers. His powers are broadly similar to what we see in the movie as well.

*Danielle Moonstar*: has a lot of similarities to the movie version – her father was killed by a demonic bear, and she is a member of the Cheyenne native Americans. She was rescued from the Hellfire Club and joined the New Mutants, and they eventually confronted and stopped the demon bear, in the process getting her parents back. Dani decided to stay with the New Mutants, however. Her original codename was Psyche, but she later changed this to Mirage. Her powers are slightly greater in the comics, able to project illusions, but also fire psionic bolts (and use a psionic bow and arrow). When she went to Asgard, she discovered she was also part of the Valkyrior, which gave her another set of super powers, akin to Valkyrie's. She remained in Asgard for a time, but later returned to Earth to join X-Force, and then the X-Men. She lost her powers when Scarlet Witch removed mutants from the Earth, though got them back again and rejoined the New Mutants when Magik reformed them to fight Legion. She has since worked alternately with the X-Men and Asgard.

*Roberto DaCosta*: is the son of wealthy businessman and Hellfire Club member Emmanuel da Costa. After being picked on by racists, Roberto – or Bobby – found his powers manifesting, which exacerbated the situation. In the comics the manifestation of his powers is different to the movie – he essentially becomes a totally black entity with solar dots when using his powers. Donald Pierce tried to kidnap him, but he was rescued by Rahne and Xi'an Coy Manh (or, more accurately, Xuân Cao Mạnh, aka Karma). He fought Sam Guthrie, but Sam changed his allegiences and the pair became friends as they joined Xavier's School. Over time, Bobby joins X-Force, is offered his father's seat at the Hellfire Club (and indeed becomes Lord Imperial of the Club), rejoins the New Mutants, joins the Avengers, buys AIM (turning it into Avengers Idea Mechanics) and is later killed when he destroys an amulet allowing Magik to properly use her powers. Naturally he is resurrected…

*Cecilia Reyes*: was created by Scott Lobdell and Carlos Pacheco for *X-Men Vol 2 #65 (Jun, '97)*, where she joined the X-Men on a very temporary basis. She has the ability to create a forcefield around herself, but prefers to remain a doctor, rather than act as a superhero. As such, both her personality and appearance are very different to the Reyes we see in the film. She has most recently been seen on Krakoa, Magneto's island home for mutants.

*And:* Rahne's father, Reverend Craig Sinclair, was a member of the Purifiers – a religious group started by Reverend William Stryker. He appears in *Marvel Graphic Novel #4: The New Mutants* for the first time, created by Claremont and McLeod, and was later killed by his own daughter (ironically because he had her brainwashed).

**Ratings:** *IMDB:* 68%; *Rotten Tomatoes:* 28%; *Metacritic:* 35

**Review: New Mutants** tries very hard to be something different, and almost succeeds. The idea of a horror film with superheroes is a great one, and the New Mutants are a good team to explore that idea. However, the revelation of the villain is weak (it clearly feels as though that was changed somewhere along the way), and the reluctance to acknowledge their place in the X-Men universe mean the New Mutants are always hinting at something that isn't really a secret. However, the casting works well, and the characters are handled in a fashion true to the source material. All in all, this is a better closing chapter for the X-Men than **Dark Phoenix** was.

## The Victims

The somewhat surprising acquisition of Fox by Disney caused a lot of upheaval, with both **Dark Phoenix** and **New Mutants** looking as though, despite having release dates and production completed, they would be pulled from release. Instead they were both delayed by some time, but avoided the dreaded straight-to-DVD.

However, there were other films that had been talked about with a degree of pre-production having started.

**Doom** was a spin-off film from *The Fantastic Four* that would have centred on Latverian dictator Victor Von Doom. The film was announced in 2017, with Noah Hawley attached to write the script which, by 2018, he confirmed had been completed. Hawley claims the movie was a geopolitical thriller in which Doom would put a dome over Latveria and then communicate with the outside world via a journalist. In March, 2019 he met with Kevin Feige, and suggested that Feige wasn't overly interested in pursuing the idea.

**X-Force** should have begun filming in October, 2018 as one of three "X" films for 2019. Both Ryan Reynolds and Josh Brolin were signed on to reprise Deadpool and Cable, while their support ranged from a number of characters both from **Deadpool 2** and **New Mutants**. Needless to say the filming didn't take place, and the movie has been sent the way of all things.

One of the other films scheduled for 2019 was **Gambit** with a release date of February, though this was later pushed back to March, 2020. It was hoped that Taylor Kitsch would spin-off from **X-Men Origins: Wolverine**, but the failure of that film killed the idea, until 2014 when Channing Tatum expressed an interest in playing the part. Josh Zetumer was hired in 2015 to write, and October 2016 became the first release date, though it was thought the character might appear in **X-Men: Apocalypse**. Then things started to go awry. Rupert Wyatt was hired as director in June, 2015, casting Lea Seydoux as Belladonna, but a month later and Tatum suddenly seemed hesitant about the project. By September he was back on board, but Wyatt was gone after his budget was slashed thanks to **Fantastic Four**. Doug Liman became the new director in November, with filming set for March, 2016 on a script now written by Tatum and Reid Carolin. This was pushed back to the end of 2016, and Seydoux was let go. In August, 2016 Liman was gone as well, unhappy with the script. By May, 2017 the project seemed in limbo, though Tatum was still keen. In October Gore Verbinski became the new director, but in January, 2018 he had gone as well, due to scheduling conflicts. Or creative differences. At this point, producer Simon Kinberg kept pushing back production and release dates, while assuring everyone the movie was in active development and the Disney buyout wouldn't have an effect. He was clearly wrong.

## VENOM: LET THERE BE CARNAGE
### *(5/10/2021)*

As Eddie and Venom attempt to reconcile their difficult lifestyle, Cletus Kassidy summons Brock and inadvertently gives away the location of his many victims. Furious Kassady bites Brock before his execution and this gives him part of the symbiote, spawning Carnage. After recovering his girlfriend, the mutant Shriek, Kassady and Shriek decide to marry with the three claiming Brock,

Mulligan and Venom to be their victims in attendance.

**Cast:** *Tom Hardy (Eddie Brock/Venom), Michelle Williams (Anne Weying), Naomie Harris (Frances Barrison/Shriek), Reid Scott (Dr Dan Lewis), Stephen Graham (Patrick Mulligan), Woody Harrelson (Cletus Kassady)*

**Dir:** Andy Serkis; **Writers:** Kelly Marcel, Tom Hardy; **Prod:** Avi Arad, Matt Tolmach, Amy Pascal, Kelly Marcel, Tom Hardy, Hutch Parker; **Music:** Marco Beltrami, **Exec.Prod:** Jonathan Cavendish, Ruben Fleischer, Barry Waldman; **DOP:** Robert Richardson; **Prod.Des.:** Oliver Scholl; **Ed:** Maryann Brandon, Stan Salfas; **Costumes:** Joanna Eatwell; **Colombia Pictures/Arad Productions/Matt Tolmach Productions/Pascal Pictures/Marvel Entertainment**; 97; $110m; $507m

**Stan Spotting:** He is on the cover of some of the magazines in Mrs Chen's shop.

**Notes:** Andy Serkis wanted to film the movie in the UK, and complete post production there as well, which happened to work in with Sony's plans. The film features appearances by both Tom Holland as Peter Parker/Spider-Man, and J K Simmons as J Jonah Jameson (though neither are credited) in a post-credits sequence that shows they have been transported into the MCU. This was not filmed for this movie, but rather is lifted from **Spider-Man: Far From Home**. The film was originally scheduled for release in October, 2020, but was delayed because of the COVID-19 pandemic. Filming on **The Matrix Resurrections** created some problems for filming when helicopters from the other production were in view on the shoot, but this was incorporated into the story. Peggy Lu returns to play Mrs Chen in the film.

**Comic Notes:** *Shriek*: or Frances Barrison, looks much the same as her comic appearance (although she is white in the comics), and is usually the lover of Cletus Kasady. She first appeared in *Spider-Man Unlimited #1 (May, '93)*, created by Tom DeFalco and Ron Lim, and had a very unhappy upbringing, abused by her mother. She was initially named Sandra Deel, but this was retconned so it was just a pseudonym for Barrison. After an encounter with Cloak, her mutant abilities were awakened and she gained the power to manipulate sound. When Carnage escaped Ravencroft, he took Shriek with him, and they then formed a crime family. She has been a member of a number of crime syndicates, but always ends up working with Carnage at some point. Most recently she and her lover formed a cult to worship Knull, the power behind Gorr the God Butcher.

*Patrick Mulligan*: first appeared in *Venom/Carnage #1 (Sep, '04)*, created by Peter Milligan and Clayton Crain. He's broadly the same as the comics version – though he's an NYPD cop in the comics – however in the comics he becomes the host to Carnage's symbiote child, Toxin. However, Mulligan's dedication to law and order saw Toxin actually have a strong moral core, and this brought him into conflict with both Carnage and Venom. Mulligan managed to tame Toxin completely, but he was later killed by Blackheart. The Toxin symbiote then bonded with Eddie Brock, before finding Bren Waters and bonding more permanently with him.

**Ratings:** *IMDB:* 68%; *Rotten Tomatoes:* 28%; *Metacritic:* 35

**Review: Venom** wasn't great, but its sequel is a serious step down in quality. Perhaps what's worse, though, is the film seems to be building towards Eddie and Venom crossing into the MCU, rather than concentrating on a story that pits Venom against Carnage. The two only really come into contact with each other at the end of the film, which is disappointing, as there should be a lot more interaction between the two enemies. Sadly, like the first film, this seems a wasted opportunity, in particular Carnage, who had the potential to be so much more than what we get in this.

## MORBIUS
### *(10/3/2022)*

Michael Morbius works hard to cure the affliction that he has struggled with since youth, but in the process gives himself powers akin to a vampire, along with a disturbing lust for blood. As he tries to keep his new powers under control, Morbius becomes the target of an FBI agent named Simon Stroud, investigating Morbius' kills. However, a new victim puzzles Morbius who realises that his surrogate brother – a man suffering the same affliction as himself – may have taken Morbius' cure and been cursed in the same way.

**Cast:** *Jared Leto (Dr Michael Morbius),* Matt Smith (Lucien), *Adria Arjona (Dr Martine Bancroft), Jared Harris (Dr Emil Nicholas),* Al Madrigal (Alberto "Al" Rodriguez), *Tyrese Gibson (Simon Stroud)*

**Dir:** Daniel Espinosa; **Writers:** Matt Sazama, Burk Sharpless; **Prod:** Avi Arad, Matt Tolmach, Lucas Foster; **Music:** Jon Ekstrand, **Exec.Prod:** Jared Leto, Emma Ludbrook, Louise Rosner; **DOP:** Oliver Wood; **Prod.Des.:** Stefania Cella; **Ed:** Pietro Scalia; **Costumes:** Cindy Evans; **Colombia Pictures/Arad**

**Productions/Matt Tolmach Productions/ Marvel Entertainment**; 104; c $80m; $167m

**Notes:** Morbius was intended to appear in **Blade** as a cameo, before essentially being the villain of **Blade II**, but this idea was ultimately dropped. At one point, Jon Watts looked at both Blade and Morbius to appear in **Spider-Man: Far From Home**. Again, this didn't happen, but it led to the development of a Morbius movie, with Jared Leto attached. Matt Smith joined the cast after being convinced by Karen Gillan to do a Marvel movie. Michael Keaton was brought in to reprise the role of Adrian Toomes/Vulture, and the first trailer featured Keaton as well as a strong reference to Spider-Man. All of this was dropped for the movie, and Keaton's scenes were reshot, and changed into the post-credit sequences. J K Simmons also filmed sequences as J Jonah Jameson, but these were all deleted. The film was a box office disaster, but when a number of memes appeared online, Sony re-released the film, thinking it had found a new audience. It had not. Leto's method acting meant he used crutches off screen, causing problems when he needed to go to the toilet. Tyrese Gibson spoke about his character's cyborg arm, but this never appears in the movie, suggesting that any scenes relating to the arm were also deleted.

**Comic Notes:** *Michael Morbius MD PhD*: made his first appearance in *The Amazing Spider-Man #101 (Oct, '71)* where he was created as a new villain for Spider-Man. The intention, like other Spider-Man villains, was to create a character who was sympathetic, and as such they came up with a man who suffered from a rare blood disease that is ultimately fatal. He was brilliant, and channelled that genius into finding a cure for his illness, winning a Nobel Prize along the way. He worked with his friend Emil Nikos, and his assistant Martine Bancroft, who he would then fall in love with. The cure he comes up with backfires and turns him into a vampire and after meeting Curt Connors (who's Lizard mutation had become worse), he was able to find a cure for Connors, and also for people who had started to mutate as a result of Morbius himself attacking them. This cure did not work on Morbius, and when Morbius attacked Martine, he fled the city. He would go onto encounter a number of Marvel's more monstrous denizens, including Man-Thing, Man-Wolf, Ghost Rider, Jack Russell and Tara the Girlchild, while fighting the X-Men, the Fantastic Four, Blade and often Spider-Man. Morbius association with CIA Agent Stroud saw him become less the villain and more an anti-hero. After joining the Legion of Monsters, he once again came into conflict with Spider-Man and was incarcerated in The Raft, which cause him to suffer mental depression. He was most recently seen encountering Nikos' son Christos.

 *Martine Bancroft*: is physically different in the comics, where she's white with blonde hair, though she does have red eyes. There are a lot of broad character similarities, though – she is Morbius' lab assistant and fiancée in the comics, and was killed by David Langford, before Lilith resurrected her. She has been turned vampiric on two occasions, the latter time in order to become a true vampire and spend eternity with Morbius, though she could not control the vampire side of her. Morbius killed her a second time in order to save Spider-Man. She first appeared in *The Amazing Spider-Man #102 (Nov, '71)*, created by Roy Thomas and Gil Kane.

*Emil Nikos*: is the comic version of Emil Nicholas, and the character of Milo is also loosely based on Nikos as well. Like Bancroft, he first appeared in *The Amazing Spider-Man #102* (created by Thomas and Kane), and is Morbius' best friend, having gone to university together. He was working with Morbius to cure his blood disease, but was actually Morbius' first victim when the cure failed. Baron Blood later transformed him into a vampire, in which form he remains to this day.

*Simon Stroud*: first appeared in *Creatures on the Loose #30 (Apr, '74)*, created by Doug Moench and George Tuska. Physically he's a white man with blonde hair, so differs significantly from the movie version, but he also worked for the CIA rather than the FBI, before ending up in NYPD. He's good friends with the Black Widow, and worked with Spider-Man to hunt both Morbius and the Man-Wolf.

**Ratings:** *IMDB:* 52%; *Rotten Tomatoes:* 16%; *Metacritic:* 35

**Review:** This film is an absolute disaster. Leto's performance is misjudged from the outset, and Morbius lacks any interesting depth. Matt Smith embraces the stupidity of the film, and seems to be having a fun time, but the rest of the cast have no idea what sort of film this is supposed to be, and as such there is no consistency. All of which is indicative of some terrible direction. By the time you get to the post-credit sequences, the film has lost any goodwill the audience had, and so the Vulture scenes that are jammed into the film lose any momentum, though they also make no sense either. A genuinely awful Marvel film.

*Coming Soon…*

## MADAME WEB
### *(16/2/2024)*

**Cast:** *Dakota Johnson (Cassandra Webb/Madame Web), Sydney Sweeney (Julia Carpenter/Spider-Woman),* Celeste O'Connor, Isabela Merced, Tahar Rahim, Emma Roberts, Mike Epps, Adam Scott, Zosia Mamet

**Dir:** S J Clarkson; **Writers:** Burk Sharpless, Matt Sazama; **Prod:** Lorenzo di Bonaventura, Erik Howsam, Palak Patel; **DOP:** Mauro Fiore; **Colombia Pictures/Marvel Entertainment/Di Bonaventura Pictures**

## VENOM III
### *(12/7/2024)*

**Cast:** *Tom Hardy (Eddie Brock/Venom),* Juno Temple, Chiwetel Ejiofor

**Dir:** Kelly Marcel; **Writers:** Kelly Marcel, Tom Hardy; **Prod:** Avi Arad, Matt Tolmach, Amy Pascal, Kelly Marcel, Tom Hardy, Hutch Parker; **DOP:** Fabian Wagner; **Colombia Pictures/Marvel Entertainment/Arad Productions/ Pascal Pictures/Hutch Parker Entertainment/Matt Tolmach Productions**

## KRAVEN THE HUNTER
### *(30/8/2024)*

**Cast:** *Aaron Taylor-Johnson (Sergei Kravinoff/Kraven the Hunter), Ariana DeBose (Calypso),* Russell Crowe (Kravinoff Sr), *Fred Hechinger (Dmitri Smerdyakov/Chameleon), Alessandro Nivola (Aleksei Sytsevich/Rhino), Christopher Abbott (Foreigner)*

**Dir:** J C Chandor; **Writers:** Art Marcum, Matt Holloway, Richard Wenk; **Prod:** Avi Arad, Matt Tolmach, David Householter; **DOP:** Ben Davis; **Ed.:** Craig Wood; **Colombia Pictures/Marvel Entertainment/Arad Productions/Matt Tolmach Productions**

# II. TELEVISION FILMS

## SPIDER-MAN
### (14/9/1977 - CBS)

Part time student and photographer, Peter Parker is bitten by a radioactive spider which gives him astonishing powers. Cult leader Edward Byron has the ability to hypnotise his followers into doing exactly what he wants, and with his desire to acquire $50 million, Byron comes up with a fiendish way to blackmail the entire city.

**Cast:** *Nicholas Hammond (Peter Parker/Spider-Man), David White (J Jonah Jameson),* Michael Pataki (Captain Barbera), *Hilly Hicks (Joe "Robbie" Robertson),* Lisa Eibacher (Judy Tyler), Dick Balduzzi (Delivery Man), *Jeff Donnell (May Parker),* Robert Hastings (Monahan), Barry Cutler (Purse Snatcher), Thayer David (Edward Byron)

**Dir:** E W Swackhamer; **Writer:** Alvin Boretz; **Prod:** Edward J Montagne; **Music:** Johnnie Spence; **Exec.Prod:** Charles W Fries, Daniel R Goodman; **DOP:** Fred Jackman; **Prod.Des.:** William H Tuntke; **Ed:** Aaron Stell; **Costumes:** Frank Novak; **Charles Fries Productions/Dan Goodman Productions/Danchuk Productions**; 88

**Notes:** A telemovie that ultimately led to being a backdoor pilot for a television series proper. It generated enough interest not just to get the television series, but also to get a limited theatrical release.

**Ratings:** *IMDB:* 65%

**Review:** For a pilot episode it's not particularly dreadful. The cast does its job effectively, but no one is astonishingly good in their roles (though Hilly Hicks delivers a unique Robbie Robertson), and the entire seventies flavour of the production can induce a laugh out of context. Unfortunately, while the Spider-Man side of things is quite good, it's the actual villainous plot that tends to be a bit of a bore, and the resolution is pretty predictable.

## DR STRANGE
### (6/9/1978 - CBS)

The Nameless One sends Morgan Le Fay back to Earth to atone for her failure to kill the sorcerer Lindmer, but this time Lindmer has an unwitting apprentice in the wings – psychiatrist Stephen Strange. When Le Fay uses Clea Lake in her attempt to kill Lindmer, Strange's psychic bond with Lake drags him into Lindmer's world.

**Cast:** *Peter Hooten (Dr Stephen Strange), Clyde Kusatsu (Wong), Jessica Walter (Morgan Le Fay), Eddie Benton (Clea Lake),* Philip Sterling (Dr Frank Taylor) and John Mills as Lindmer, June Barrett (Sarah), Sarah Rush (Nurse)

**Dir/Writer/Exec.Prod:** Philip DeGuere; **Prod:** Alex Beaton; **Music:** Paul Chihara; **DOP:** Enzo A Martinelli ASC; **Prod.Des.:** William H Tuntke; **Ed:** Christopher Nelson; **Costumes:** Yvonne Wood; **CBS/Universal/MCA Television**; 93

**Notes:** A backdoor pilot for a television series; as such Hooten gets a "starring" credit and Kusatsu an "also starring" credit, suggesting they would have been the two regular characters in the series. Eddie Benton later changed her name to Anne-Marie Martin. Stan Lee worked as a consultant on the movie. Although never stated as such, Lindmer essentially performs the function of the Ancient One in the comics.

**Comic Notes:** *Stephen Vincent Strange & Wong:* made their first appearance in the comics in *Strange Tales #110 (Jul, '63),* both created by Stan Lee and Steve Ditko. Strange's comic backstory is quite different to the telemovie; in the comics, he is an arrogant surgeon who has the bones in his hands crushed in a car accident, and after much searching, and becoming homeless in the process, Strange finds the Ancient One – who is sorcerer supreme of our world. When he stops Baron Mordo killing an old man, he is taught the mystic arts. It turns out the Ancient One knew of Mordo's treachery and was using it as a test for Strange. Wong, is also a little different to his appearance in the comics. He comes from a long line of Chinese monks who have vowed to stand by and serve the reigning sorcerer supreme. Wong was sent by the Ancient One to serve Strange for when he becomes the next sorcerer supreme, something the man has done loyally for many years. An expert martial artist, he has over time been romantically involved with Strange's secretary Sara Wolfe, and alongside her been made co-director of the Stephen Strange Memorial Metaphyhsical Institute (when they all thought Strange was dead.)

    *Clea Lake:* turned up for the first time in *Strange Tales #126 (Nov, '64),* also created by Lee and Ditko. Again her backstory is considerably different – she is the sorceress supreme of the Dark Dimension, and, impressed by the courage of Stephen Strange, stood by him against the demonic Dormammu. She would later marry Strange, while her mother Umar fell in love with Mordo, and the two couples have fought over control of the Dark Dimension. Clea left the Dark Dimension to live with her husband, but later returned to lead the resistance there.

*Morgan Le Fay*: first appeared in *Black Knight #1 (May, '55)* and is based on the character from Arthurian Legend, though this version was created by Lee and Joe Maneely. Originally she was a recurring foe for the Black Knight, but she would go onto appear in *Spider-Woman #2 (May, '78)* where Marv Wolfman and Carmine Infantino reimagined her as attempting to obtain the Darkhold (the Book of Sins). She has fought Spider-Woman and the Avengers, and been in a relationship with Doctor Doom, though he is responsible for where she is now – 1,000,000 BC. Curiously, in the comics, her first meeting with Dr Strange was in *The Avengers #240 (Feb, '84)* – six years after this film.

**Ratings:** *IMDB:* 55%; *Rotten Tomatoes:* 52%

**Review:** Given this was made in the seventies it has the lack of pace and poor visual effects of that period (though the Nameless One is particularly bad). However if you can get past those, this pilot isn't terrible, and John Mills is – as usual – outstanding. Chihara's soundtrack sounds like it should be on a John Carpenter film, which is no bad thing. The biggest problem with the movie is that for virtually the entire film, Strange never uses magic. In fact he only seems to use it once at the very end for humour's sake. But there's something to enjoy in this film, and for the most part the acting is really quite good, and it's a great set up for the possible series. More importantly it's a brave attempt to take a very different Marvel character and give him a somewhat decent live action outing.


## CAPTAIN AMERICA
### *(19/1/1979 - CBS)*

Steve Rogers, the son of a government agent, whose patriotism has earned him the nickname Captain America, is involved in a car accident and almost fatally wounded. Injected with the experimental FLAG serum, the former marine is approached by Dr Mills – a colleague of his father's – to start life a superhero. When Rogers becomes the target of assassins, he is forced to agree.

**Cast:** *Reb Brown (Steve Rogers/Captain America),* Len Birman (Dr Simon Mills), Heather Menzies (Dr Wendy Day), Robin Mattson (Tina Hayden), Joseph Ruskin (Rudy Sandrini), Lance Le Gault (Harley), Frank Marth

(Charles Barber), Steve Forrest (Lou Brackett), Chip Johnson (Jerry), James Ingersoll (Lester Wiant)

**Dir:** Rod Holcomb; **Writers:** Don Ingalls, Chester Krumholz; **Super.Prod:** Daniel McPhee; **Music:** Lance Rubin; **Exec.Prod:** Allan Balter; **DOP:** Ronald W Browne; **Prod.Des.:** Lou Montejano; **Ed:** Michael S Murphy; **Costumes:** Charles Waldo; **Universal Television**; 97

**Notes:** An unusual adaptation of the original source material that clearly changes Rogers' backstory considerably, and doesn't use any elements from the comics aside from Rogers himself. Whilst not a critical success, it was financially enough to earn a sequel.

**Ratings:** *IMDB:* 37%

**Review:** Even accounting for context, this movie is pretty awful. The fight sequences have the same terrible choreography that the ***Spider-Man*** ones do, but in this case, Captain America himself is terrible and so there's very little to like. And, it's difficult to get past the perspex shield. One of the most awful movies to be associated with Marvel. Though, it could be worse…

## CAPTAIN AMERICA II: DEATH TOO SOON
### *(23 & 24/11/1979 - CBS)*
While Captain America deals with a social security fraud, General Miguel kidnaps Professor Ilson, and forces him to complete his aging serum (and antidote). With the chemicals completed, Miguel heads for Oregan, where he plans to hold Portland ransom for one million dollars! Until Captain America arrives to stop him.

**Cast:** *Reb Brown (Steve Rogers/Captain America),* Len Birman (Dr Simon Mills), Connie Sellecca (Dr Wendy Day), Christopher Lee (Miguel), Katherine Justice (Helen Moore), Christopher Cary (Professor Ilson), Bill Lucking (Stader), Stanley Kamel (Kramer), Ken Swofford (Everett Bliss), Lana Wood (Yolanda)

**Dir:** Ivan Nagy; **Writers:** Wilton Schiller, Patricia Payne; **Super.Prod:** Daniel McPhee; **Music:** Peter Carpenter, Mike Post; **Exec.Prod:** Allan Balter; **DOP:** Vincent A Martinelli; **Prod.Des.:** David L Snyder; **Ed:** Michael S Murphy; **Costumes:** Yvonne Wood; **Universal Television**; 88

**Notes:** Pushing the definition of movie, this telemovie was actually broadcast in two parts, over two consecutive nights. This was more a pilot for a television series, though the response to this telemovie killed the idea. Surprisingly it did get a theatrical release in some parts of the world.

**Ratings:** *IMDB:* 39%

**Review:** Abominably bad. Not even Christopher Lee can save this mess.

## THE INCREDIBLE HULK RETURNS
### *(22/5/1988 - NBC)*

Banner's life has changed considerably in the last two years, and he has not changed into the Hulk during that time. However, he meets a former student, Donald Blake, who reveals that he has come into possession of an enchanted hammer that contains the soul of Thor, who must now serve him. Meanwhile, thieves take a gamma transponder Banner has been working on which may cure him of the Hulk.

**Cast:** *Bill Bixby (David Banner), Lou Ferrigno (Hulk),* Jack Colvin (Jack McGee), Lee Purcell (Maggie Shaw), Charles Napier (Mike Fouche), John Gabriel (Joshua Lambert), Jay Baker (Zack Lambert), Tim Thomerson (Jack LeBeau), *Eric Kramer (Thor), Steve Levitt (Donald Blake),* William Riley (Sgt Lindsey), Tom Finnegan (Capt Brills), Donald Willis (Elwood), Carl Nick Ciafalio (Barner)

**Dir/Writer:** Nicholas Corea; **Super.Prod:** Daniel McPhee; **Music:** Lance Rubin; **Exec.Prod:** Bill Bixby, Nicholas Corea; **DOP:** Chuck Colwell; **Prod.Des.:** Michael Parker; **Ed:** Janet Ashikaga, Briana London; **Costumes:** John Casey; **Bixby-Brandon Productions/New World Television**; 100

**Notes:** Essentially a reunion movie, CBS and Universal who made the television series were not interested in the production, and so Bixby's own production company, along with New World made the movie, with NBC broadcasting it. It exceeded expectations ratings wise. Colvin suffered a stroke not long after the making of the movie, and retired from acting. There was hope that a Thor television series might come from the movie, but it was decided the success of the film was the Hulk rather than Thor, and so a second movie was commissioned and the series was dropped. Oddly, this is the first time in the run of *The Incredible Hulk*, that other Marvel elements made an appearance.

**Comic Notes:** *Thor:* first appeared in *Journey into Mystery #83 (Aug, '62)*, created by Stan Lee, Larry Lieber and Jack Kirby, Marvel's own gods. His origin story is has changed over time in the comics, but this movie uses the elements of the original. There he was a man named Donald Blake who finds a walking stick and, when he strikes it, discovers the stick is Mjolnir and he turns into Thor. This was retconned quite quickly so that Thor had been banished to Earth by Odin, to the body of Blake as punishment, and then retconned again so that Blake had always been Thor, just unaware of it. He was a founding member of the Avengers from early on as well. Thor's powers are mostly the same in both versions.

**Ratings:** *IMDB:* 57%

## THE TRIAL OF THE INCREDIBLE HULK
### *(7/5/1989 - CBS)*

Heading north to a big city, Banner adopts a new identity, but becomes the Hulk when he sees a girl being molested, and as Banner is subsequently arrested for the crime. He is represented by lawyer Matt Murdock, but soon falls foul of a criminal mastermind named Wilson Fisk. Hulk finds himself getting help from the masked superhero Daredevil, as he takes on Fisk.

**Cast:** *Bill Bixby (David Banner), Lou Ferrigno (Hulk),* Marta Dubois (Ellie Mendez), Nancy Everhard (Christa Klein), Nicholas Hormann (Edgar), Richard Cummings Jr (Al Pettiman), Joseph Mascolo (Albert G Tendelli), *John Rhys-Davies (Wilson Fisk), Rex Smith (Matt Murdock/Daredevil),* Linda Darlow (Fake Nurse), John Novak (Denny), Dwight Koss (John), *Mark Acheson (Turk)*

**Dir:** Bill Bixby; **Writer:** Gerald Dipego; **Prod:** Robert Ewing, Hugh Spencer-Phillips; **Music:** Lance Rubin; **Exec.Prod:** Bill Bixby, Gerald Dipego; **DOP:** Chuck Colwell; **Prod.Des.:** Doug Higgins; **Ed:** Janet Ashikaga; **Costumes:** Judy Truchan, Jori Woodman; **Bixby-Brandon Productions/New World Television**; 95

**Notes:** Again there was some hope that this might lead to a Daredevil television series, but again it did not eventuate, though a third Hulk film was commissioned. This is the only time in *The Incredible Hulk* run where the Hulk actually wears his traditional purple pants. The decision was made to give

Daredevil a black costume, something which the comics later adopted briefly, and something which Stan Lee was less than impressed with.

**Stan Spotting:** The foreman of the jury in Banner's nightmare.

**Ratings:** *IMDB:* 59%

## THE DEATH OF THE INCREDIBLE HULK
### *(18/2/1990 - CBS)*

Banner travels to Portland where he believes Dr Ronald Pratt may be able to cure him, and secretly guides Pratt's work. Meanwhile a Russian spy named Jasmin is sent to retrieve Pratt's work or her sister Bella will be killed. Banner comes clean to Pratt, and the pair work on the cure together with his wife Amy, but when Amy is kidnapped, Banner becomes the Hulk to stop Jasmin's superiors.

**Cast:** *Bill Bixby (David Banner), Lou Ferrigno (Hulk),* Elizabeth Gracin (Jasmin), Philip Sterling (Dr Ronald Pratt), Barbara Tarbuck (Amy Pratt), Anna Katerina (Bella), John Novak (Zed), Andreas Katsulas (Kasha), Chiltron Crane (Betty), Carla Ferrigno (Bank Teller), Duncan Fraser (Tom), Dwight Mcfee (Brenn), Lindsay Bourne (Crane), Mina E Mina (Pauley), Marlane O'Brien (Luanne Cole)

**Dir:** Bill Bixby; **Writer:** Gerald Dipego; **Prod:** Robert Ewing, Hugh Spencer-Phillips; **Music:** Lance Rubin; **Exec.Prod:** Bill Bixby; **DOP:** Chuck Colwell; **Prod.Des.:** Doug Higgins; **Ed:** Janet Ashikaga; **Costumes:** Trish Keating; **Bixby-Brandon Productions/New World Television**; 95

**Notes:** For a long time it was intended that She-Hulk would appear in this film, with the intention to do a fourth that would have introduced Iron Man, but both these ideas failed to materialise. Even though Banner dies, there were firm plans for a sequel "The Revenge of the Incredible Hulk". Popular myth has it that Bill Bixby's declining health stopped the sequel, though equally popular myth suggests that Banner wouldn't have appeared at all; in fact it would have been the Hulk with Banner's mind. The truth, however, is that the ratings had declined to the point where a fourth film was not commissioned.

**Ratings:** *IMDB:* 58%

# GENERATION X

## *(20/2/1996 - Fox)*

Emma Frost and Sean Cassidy recruit Jubilee and Skin to the Xavier School for Gifted Youngsters, where the pair join the other four students – all mutants, all escaping the registration act. However, Dr Tresh is seeking to find the X-Factor which gives mutants their abilities, and as Jubilee and Skin discover a device Tresh used to enter the astral plane, Tresh is already there waiting for them.

**Cast:** Matt Frewer (Dr Russell Tresh), *Finola Hughes (Emma Frost), Jeremy Ratchford (Sean Cassidy), Heather McComb (Jubilation "Jubilee" Lee), Austin Rodriguez (Angelo "Skin" Espinosa), Amarilis (Monet St Croix/M), Bumper Robinson (Mondo),* Suzanne Davis (Arlee Hicks), Randall Slavin (Kurt Pastorius), Kevin McNulty (Ralston), Lalainia Lindbjerg (Kayla), Garry Chalk (Det Gaines), Lynda Boyd (Alicia Lee), Wally Dalton (Pruitt), Kavan Smith (Lance), Noel Greer (Bruce), Ken Ryan (Dr Carlysle)

**Dir:** Jack Sholder; **Writer:** Eric Blakeney; **Prod:** David Roessell; **Music:** J Peter Robinson; **Exec.Prod:** Avi Arad, Eric Blakeney, Matthew Edelman, Bruce Sallan, Stan Lee; **DOP:** Bryan England; **Prod.Des:** Douglas Higgins; **Ed:** Michael Schweitzer; **Costumes:** Maya Mani; **New World Entertainment/ Marvel Entertainment Group**; 87

**Notes:** The film was a backdoor pilot for a television series that never got greenlit. Hatley Castle is used to represent Xavier's School – the same location as used for the first three X-Men films. In early drafts, Dazzler and Boomer were the lead characters. Jubilee was added as she was popular in the comics as a member of *Generation X.* Husk and Chamber were also set to appear, but were replaced by Arlee and Kurt as the formers' powers would have been too difficult to create on screen. Ratchford also voiced Banshee in the animated series. The British version includes some swearing and a scene of Jubilee stripping that was removed from the US edit.

**Comic Notes:** *Skin, Mondo and M:* Skin first appeared in *The Uncanny X-Men #317 (Oct, '94)*, and was created by Scott Lobdell and Joe Madureira. M first appeared in *The Uncanny X-Men #316 (Sep, '94).* Mondo first appeared in *Generation X #3 (Jan, '95)* – though this was actually a clone, and the real Mondo finally turned up in *#60 (Feb, '00)!* Both were created by Scott Lobdell and Chris Bachalo. All three's powers are represented fairly accurately; as indeed are Emma Frost's and Sean Cassidy's. The appearance of several is quite different – in the comics Jubilee is East Asian, Skin's skin is grey and

Mondo is Caucasian, though the names of the characters are all correct (Mondo's real name has never been revealed). M, Mondo, Skin and Jubilee were indeed all members of Generation X (well, the cloned Mondo was – when the real version turned up, he was actually loyal to Black Tom Cassidy, and part of Cassidy's plan to get revenge on his cousin, Sean), and they were at Xavier's when Cassidy and Frost were the headmasters. A surprisingly faithful rendition.

**Ratings:** *IMDB:* 48%; *Rotten Tomatoes:* 52%

**Review:** An oddly inconsistent film that doesn't particularly play to the strengths of any part of the production, but in spite of that feels like there is something good trying to get out, particularly towards the end. Ratchford's accent is risible and Frewer is particularly obnoxious – acting as though he were trying to out-Carrey Jim Carrey's Riddler – but the students become quite likable towards the end. The additional UK scenes seem bizarrely out of the place (Jubilee's swearing is the only time it happens in the movie). Not a great moment for Marvel, but still with a spark of something in it.


## NICK FURY: AGENT OF SHIELD
### *(26/5/1998 - Fox)*

Reanimated by his children, Baron Von Strucker has created the organisation Hydra, and plans to unleash the Death's Head virus if he isn't paid $1 billion. With no choice SHIELD turn to their former agent, Nick Fury, in the hope he will come out of retirement and take on his old nemesis one last time.

**Cast:** *David Hasselhoff (Nick Fury), Lisa Rinna (Countess Valentina de Fontaine), Sandra Hess (Andrea Von Strucker/Viper), Neil Roberts (Alexander Pierce), Garry Chalk ("Dum Dum" Dugan), Tracy Waterhouse (Kate Neville),* Tom McBeath (Director General Pincer) and *Ron Canada (Gabe Jones), Pater Haworth (Arnim Zola), Scott Heindl (Werner Von Strucker), Adrian Hughes (Clay Quartermain), Campbell Lane (Baron Wolfgang von Strucker),* Terry David Mulligan (The President)

**Dir:** Ron Hardy; **Writer:** David Goyer; **Prod:** David Roessell; **Music:** Kevin Kiner; **Exec.Prod:** Avi Arad, Stan Lee, Tarquin Gotch, Bob Lemchen; **DOP:** James Bartle; **Prod.Des:** Douglas Higgins; **Ed:** Drake Silliman ACE; **Costumes:** Monique Prudhomme; **Fury Productions Limited Partnership/ National Studios Inc/20th Century Fox Television/Marvel Enterprises**; 90

**Notes:** A strange film that was surprisingly not a back door pilot. Despite the efforts to bring Hasselhoff into line with his character, there is very little connection to the comics from which the inspiration came.

**Comic Notes:** *Colonel Nicholas J Fury:* was created by Lee and Jack Kirby, first appearing in *Sgt Fury and His Howling Commandos #1 (May, '63)*. The MCU has given us virtually no back story to Fury, but in the comics he led the Howling Commandos during World War II, before becoming a Cold War spy for the CIA (where he started wearing an eyepatch due to a war injury). Later he was appointed the director of SHIELD. Fury is very much an anti-hero, having been held in a stasis thanks to an Infinity formula. That formula has now depleted, and after murdering the Watcher, Fury has been forced to take the being's place, chained to the moon. In the mainstream Marvel universe he is a white man, with grey flecked brown hair, but when the Ultimate Marvel comics were created, the decision was taken to change Fury and base his appearance on Samuel L Jackson.

*"Dum Dum" Dugan:* first appeared in *Sgt Fury and His Howling Commandos #1 (May, '63)* created by Lee and Kirby. He bears very little resemblance to his on-screen appearance here – in the comics he is famous for his moustache and bowler hat. A member of the Howling Commandoes, he stayed by Fury's side and ultimately became a member of SHIELD, with a few different excuses for how he has managed to stay relatively young, despite fighting in World War II.

*Gabe Jones:* also made his first appearance in *Sgt Fury and His Howling Commandos #1* with the same creators. He was another Howling Commando, and this movie has actually bothered to try to keep the physical resemblance similar, though the characters are nothing alike in reality.

*Kate Neville:* first appeared in *Nick Fury Vs SHIELD #3 (Aug, '88)*, created by Bob Harras and Paul Neary. She's not too far removed from the character as it appears in this movie, surprisingly.

*Andrea Von Strucker:* was never Viper of course (see **The Wolverine** for that character), but Andrea made her first appearance in *Uncanny X-Men #194 (Jun, '85)* alongside her brother Andreas, and six issues later they would be revealed as the supervillain Fenris. Created by Chris Claremont and John Romita Jr, she is really nothing like the movie version. Andrea, when holding hands with her lover and brother (yes, I'm afraid so), can generate disintegration beams. The pair fought many superheroes, but rather madly took on many villains as well. Andrea was ultimately killed by Baron Zemo, though her brother forced Arnim Zola to clone her, however Bullseye later killed this clone. As Andreas was also killed by Norman Osborn, it appears that's the end of them.

*Alexander Pierce*: first appeared in *Nick Fury Vs SHIELD #3 (Aug, '88)*, and was created by Bob Harras and Paul Neary. In the comics he is kl56879-a very competent SHIELD agent who was very loyal to Fury.

*Valentina Allegra de Fontaine*: is a very old character, first appearing in *Strange Tales #159 (Aug, '67)* created by Jim Steranko. She was indeed Fury's lover and a high ranking member of SHIELD. Of late it has been revealed that she is actually a Russian spy working for Leviathan, and has adopted the mantle of Madame Hydra. Despite this, Fury still loves her and tries to save her.

*Clay Quartermain*: appeared not long after Valentina in *#163 (Dec, '67)*, also created by Steranko for his SHIELD section in *Strange Tales*. Another high-ranking SHIELD agent, he was recruited in the comics to lead the Hulkbusters (though worked alongside both Hulk and She-Hulk), and was later revealed to have been in an intimate relationship with Jessica Jones. He was killed by the insane Doc Samson.

**Ratings:** *IMDB:* 37%; *Rotten Tomatoes:* 17%

**Review:** Very much a film for fans of Hasselhoff, who, to his credit, really does make an effort with Fury and tries to bring the comic book character to life. Sadly, most of the people around him don't seem to be making the same effort, and Goyer's script is surprisingly soulless. One of the better made-for-television films, but that still isn't saying much.

# III. ANIMATED FILMS

## ULTIMATE AVENGERS: THE MOVIE
### *(21/2/2006)*

In 1945 a US soldier undergoes an experimental treatment to turn him into a super soldier. The experiment is successful, but Captain America is lost in the Arctic waters on a mission and remains missing for seventy years. When General Nick Fury recovers his body, he uses him as the core of a new team of superheroes when the Earth is threatened by the Chitauri.

**Cast:** *Justin Goss (Steve Rogers/Captain America), Grey DeLisle (Janet Pym/ The Wasp), Michael Massee (Bruce Banner), Olivia D'Abo (Natalia Romanoff/ Black Widow), Marc Worden (Tony Stark/Iron Man), Nolan North (Hank Pym/ Giant Man), David Boat (Thor), Nan McNamara (Dr Betty Ross), Andre Ware (Nick Fury), Fred Tatasciore (Hulk/Edwin Jarvis), James Arnold Taylor (Bucky),* James K Ward (Herr Kleiser), Dee Bradley Baker, Steve Blum, Keith Ferguson, Quinton Flynn, Kerrigan Mahan, Aileen Sander

**Dir:** Curt Geda, Steven E Gordon, Bob Richardson; **Writers:** Greg Johnson, Boyd Kirkland, Craig Kyle; **Prod:** Bob Richardson; **Music:** Guy Michelmore; **Exec.Prod:** Avi Arad, Craig Kyle, Eric S Rollman; **Ed:** George Rizkillah; **Lionsgate/MLG Productions 1/Marvel Studios**; 72

**Notes:** Based on *The Ultimates* comic series by Mark Millar and Bryan Hitch – though there are some changes in order to accommodate the running time. This was the first of eight animated features Lions Gate and Marvel Entertainment agreed to create.

**Ratings:** *IMDB:* 68%; *Rotten Tomatoes:* 59%

**Review:** Depending somewhat on whether you prefer the original form of the Avengers, or if you like Mark Millar's take with the Ultimates, your reaction to this movie will probably be the same. The production is flawless, the animation extremely good and the voice work above average. I've always enjoyed *The Ultimates* so I'm inclined to rate this highly, but there are those who are the reverse.

## ULTIMATE AVENGERS 2: RISE OF THE PANTHER
### *(8/8/2006)*

The battle against the Chitauri take the Avengers to Wakanda, where Kleiser kills T'Chaka, the nation's leader. As T'Challa, his son, takes over and assumes the mantle of Black Panther, protector of Wakanda, he turns to the Avengers for help in fighting the Chitauri.

**Cast:** *Justin Goss (Steve Rogers/Captain America), Grey DeLisle (Janet Pym/ The Wasp), Michael Massee (Bruce Banner), Olivia D'Abo (Natalia Romanoff/ Black Widow), Marc Worden (Tony Stark/Iron Man), Nolan North (Hank Pym/ Giant Man), David Boat (Thor), Nan McNamara (Dr Betty Ross), Andre Ware (Nick Fury), Fred Tatasciore (Hulk/Edwin Jarvis), Jeffrey D Sams (T'Challa/ Black Panther),* James K Ward (Herr Kleiser), *Dwight Schultz (Odin),* Mark Hamill (Oiler), *Dave Fennoy (T'Chaka),* Susan Dalian (Nakinda), *Kendré Berry (Young T'Challa),* Chi McBride (Chief Elder)

**Dir:** Will Meuginot, Richard Sebast, Bob Richardson; **Writers:** Greg Johnson, Boyd Kirkland, Craig Kyle; **Prod:** Bob Richardson; **Music:** Guy Michelmore; **Exec.Prod:** Avi Arad, Craig Kyle, Eric S Rollman; **Ed:** George P Rizkillah; **Lionsgate /MLG Productions 2/Marvel Studios**; 73

**Notes:** Inspiration is taken from *The Ultimate* comics, but unlike the last time, the storyline for this film is mostly original.

**Ratings:** *IMDB:* 68%; *Rotten Tomatoes:* 50%

**Review:** Another excellent entry into the series, though with the introduction of Black Panther, someone has to take a back seat, and this time round it's Hulk who is sidelined. There's plenty for everyone else, however, and this is a worthwhile sequel.

## THE INVINCIBLE IRON MAN
### *(23/1/2007)*

Tony Stark attempts to excavate an ancient temple in China that results in him being attacked by the group called the Jade Dragons, while his friend, James Rhodes, is kidnapped. With the excavation resulting in the resurrection of China's most feared Emperor, the Mandarin, Stark builds a special suit of armour in an attempt to confront and destroy the apparition.

**Cast:** *Marc Worden (Tony Stark/Iron Man),* Gwendoline Yeo (Li Mei), *Fred Tatasciore (Mandarin), Rodney Saulsberry (James Rhodes), Elisa Gabrilli (Pepper Potts), John McCook (Howard Stark),* James Sie (Wong Chu), Stephen Mendillo (Boyer), John DeMita (Drake), George Cheung, Chris

Edgerly, Paul Nakauchi, Michael Yama

**Dir:** Patrick Archibald, Jay Oliva, Frank Paur; **Writers:** Avi Arad, Boyd Kirkland, Craig Kyle; **Prod:** Frank Paur; **Music:** Guy Michelmore; **Exec.Prod:** Avi Arad, Craig Kyle, Eric S Rollman; **Ed:** Aeolan Kelly, George P Rizkillah; **Lionsgate/ MLG Productions 3/Marvel Studios/Marvel Entertainment**; 83

**Notes:** Another largely original story that uses a variety of elements from the original Marvel comics.

**Ratings:** *IMDB:* 60%; *Rotten Tomatoes:* 78%

**Review:** Not quite as good as the previous films, this one is nonetheless quite interesting, and the use of the Mandarin is very different, but not terrible.


## DOCTOR STRANGE: THE SORCERER SUPREME
### *(14/8/2007)*

Arrogant surgeon Stephen Strange loses his skills in a car accident, and heads to the mountains of Tibet in the hope of finding some peace. However there he meets the Ancient One who reveals to him that he has a destiny as the sorcerer supreme of this world, and an ancient evil – Dormammu – is returning.

**Cast:** *Bryce Johnson (Dr Stephen Strange), Paul Nakauchi (Wong), Kevin Michael Richardson (Mordo), Michael Yama (Ancient One),* Susan Spano (Dr Gina Atwater), *Jonathan Adams (Dormammu),* Fred Tatasciore (Oliver), Tara Strong (April), Josh Keaton, Phil LaMarr, Masasa Mayo

**Dir:** Patrick Archibald, Jay Oliva, Dick Sebast, Frank Paur; **Writers:** Greg Johnson, Craig Kyle; **Prod:** Frank Paur; **Music:** Guy Michelmore; **Exec.Prod:** Avi Arad, Craig Kyle, Eric S Rollman; **Ed:** Aeolan Kelly, George P Rizkillah; **Lionsgate/MLG Productions 4/ Marvel Entertainment**; 76

**Notes:** The character of April replaces Donna Strange from the comics. Dr Donald Blake makes a cameo appearance in the film – he is, of course, the alter ego of Thor.

**Ratings:** *IMDB:* 68%; *Rotten Tomatoes:* 58%

**Review:** In spite of the somewhat unknown nature of the character (or maybe because of it), this movie pulls out a lot of stops in order to be something interesting and appealing. Surprisingly darker than its predecessors, the film has a great story, and a really good cast.

## NEXT AVENGERS: HEROES OF TOMORROW
### *(2/9/2008)*

After the Avengers are killed in battle with Ultron, Tony Stark escapes with a number of children from the group and secretly raises them, whilst also training them for battle. When the Vision returns with news of Hawkeye's long lost son, James Rogers accidentally activates the Iron Avengers, giving Ultron the chance to capture Stark. Now under the care of Vision, the children locate Francis Barton, the Hulk and Thor with a view to battling Ultron and the Iron Avengers.

**Cast:** Noah Crawford (James Rogers), Breanna O'Brien (Torunn), Aidan Drummond (Henry Pym Jr), Dempsey M Pappion (Azari), Adrian Petiw (Francis Barton), *Tom Kane (Iron Man/Ultron), Fred Tatasciore (Hulk), Shawn MacDonald (Vision), Michael Adamthwaite (Thor), Ken Kramer (Bruce Banner), Nicole Oliver (Betty Ross)*

**Dir:** Jay Oliva, Gary Hartle; **Writers:** Greg Johnson, Craig Kyle, Christopher Yost; **Prod:** Gary Hartle; **Music:** Guy Michelmore; **Ed:** Ken Cravens; **MLG Productions 5/ Marvel Entertainment**; 78

**Notes:** Virtually all of this is an original story, though the concept of the Iron Avengers did first appear in the comic series *Earth X (1999-2000)*. It's unclear why so many of the characters have been recast.

**Ratings:** *IMDB:* 64%

## HULK VS
### *(20/1/2009)*

The Hulk finds himself firstly on the run from Wolverine who has been sent to bring him in, though the two must team up in order to take on some dangerous figures from their past. Then, Loki drags him to Asgard in a bid to help the god defeat his half-brother, Thor.

**Cast:** *Fred Tatasciore (Hulk), Matt Wolf (Thor), Graham McTavish (Loki), Grey DeLisle (Sif), Kari Wahlgren (Amora), Bryce Johnson (Bruce Banner), Janyse Jaud (Hela/Lady Deathstrike), Jay Brazeau (Volstagg), Jonathan Holmes (Fandral), Paul Dobson (Hogun), Michael Adamthwaite (Balder), French Tickner (Odin), Nicole Oliver (Betty Ross),* Qayam Devji (Bruce Jr), *Steve Blum (Wolverine), Mark Acheson (Sabretooth), Colin Murdock (Omega Red), Nolan North (Deadpool),* Tom Kane (The Professor), Scott McNeil, Brian Drummond, Jason Simpson

**Dir:** Sam Liu, Frank Paur, Jamie Simone; **Writers:** Christopher Yost, Frank Paur, Craig Kyle; **Prod:** Frank Paur; **Music:** Guy Michelmore; **Exec.Prod:** Kevin Feige, Eric S Rollman; **Ed:** George P Rizkillah; **Lionsgate /MLG Productions 6 Inc/Marvel Animation/Marvel Studios**; 78

**Notes:** Essentially two mini episodes, each independently made, though with the Hulk characters' casting kept the same (though not with previous releases, as again a large number of characters are recast. Quite why still remains unclear, particularly as Michael Adamthwaite who voiced Thor in **Next Avengers** has been cast in this film!). The series **Wolverine and the X-Men** exists in the same continuity as this film, and the episode "Wolverine Vs Hulk" is a direct sequel to this.

**Comic Notes:** *The Enchantress* – the more familiar name for Amora – first appeared in *Journey into Mystery #103 (Apr, '64)*, and she was created by Stan Lee and Jack Kirby. Born in Asgard, and sister to Lorelei, she learned magic from the Queen of the Norns, and discovered she had a knack for seducing others. Her character is much as it is in the movie; essentially a seductress, out to make trouble for others, especially Thor. She often works for Loki, though ironically she was killed by him during Ragnarök. Hela first appeared in the issue directly before Amora's debut - *#102 (Mar, '64)* – also by Lee and Kirby. She is the daughter of a previous Loki – the one from before an earlier Ranarök – and was appointed Goddess of Death by Odin, ruling over Hel and Niffleheim, though her ambition extends sometimes to attempting to control Valhalla as well. Again, she is similar to her movie appearance, and her motives are often somewhat opaque, but invariably self-serving.

**Ratings:** *IMDB:* 71%

**Review:** A movie of two halves in every sense of the words. The Wolverine segment is perhaps the better of the stories, with the fight a little more exacting and the reasons behind it rather nicely put together. The Thor segment is not quite as well thought out, and the Asgardians in particular come across as relatively pathetic. The voice work and animation in this section is also unimpressive.

## PLANET HULK
### *(10/2/2010)*

The Hulk wakes to find that Tony Stark and others have sent him away to prevent him causing more harm on Earth. He ultimately crashes on the planet Sakaar, where he is forced into slavery to fight gladiatorial combats. Rejecting this, Hulk forms an alliance with others and as his popularity increases, he decides to take on the person who has enslaved them; the Red King.

**Cast:** *Rick D Wasserman (Hulk), Lisa Ann Beley (Caiera), Mark Hildreth (Red King), Liam O'Brien (Hiroim), Kevin Michael Richardson (Korg), Sam Vincent (Miek), Advah Soudack (Elloe Kaifi), Michael Kopsa (Lavin Skee), Paul Dobson (Beta Ray Bill), Marc Worden (Iron Man),* Lee Tockar (Android), *Russell Roberts (Primus Vand),* Donald Adams (Governor Churik), Doug Abrahams, David Kaye, Ellen Kennedy, Cam Lane, Chantal Strand

**Dir:** Sam Liu; **Writers:** Greg Johnson, Craig Kyle, Joshua Fine; **Prod:** Frank Paur; **Music:** Guy Michelmore; **Exec.Prod:** Eric S Rollman; **Ed:** George P Rizkillah; **Lionsgate/MLG Productions 8 Inc/Marvel Animation**; 81

**Notes:** A number of characters cameo in this film, including Star Lord, Gamora and Adam Warlock. This movie is based rather directly on *The Incredible Hulk #92 - #95 (Vol 2)* "Planet Hulk" by Greg Pak and Carlo Pagulayan. Beta Ray Bill stands in for Silver Surfer, as it was unclear whether Marvel had the rights to use the character in an animated film.

**Comic Notes:** Perhaps, unsurprisingly, Caiera the Oldstrong, Angmo-Asan (better known as the Red King), Hiroim the Oldstrong and Miek first appeared in *The Incredible Hulk #92 (Apr, '06 – Volume 2)*, created by Greg Pak and Carlo Pagulayan. Elloe Kaifi, Lavin Skee and Primus Vand made their debut the following issue *#93 (May, '06 – Volume 2)*. Their characters are as close to the movie version as to be almost identical. Beta Ray Bill first appeared in *Thor #337 (Nov, '83)* created by Walt Simonson, and his backstory is somewhat different to this movie. He is an alien Korbonite who was awoken from suspended animation by Thor and after fighting, was lucky enough to separate Thor from his hammer for enough time for him to turn back into

Donald Blake. When Bill struck the cane, he became Thor, and was worthy to wield Mjolnir. Despite this, and beating Thor in a battle of Odin's design, Bill opted not to claim Mjolnir, and Odin was so impressed he created a new hammer – the Stormbreaker – and Bill took this, also gaining the powers of Thor. Since that time he has fought not only for his own people, but also alongside Asgardians and humans in a variety of battles.

**Ratings:** *IMDB:* 69%

**Review:** A really good version of the comic, with the small exception that Rick Wasserman is terrible as the voice of Hulk (again, a recasting that doesn't make any sense at all). Perhaps the only other failing is the need to rush certain aspects of the story (presumably to fit a specified running time); as such Beta Ray Bill comes across as fairly weak. Guy Michelmore's score is particularly praiseworthy.

## THOR: TALES OF ASGARD
### *(11/5/2011)*

In an effort to prove himself to his father, Thor takes Loki and stows away with the Warriors Three as they travel to Yondeheim to search for the Sword of Surtur. It is a punishing quest, and one that delves into the past as the brothers find themselves accountable for the sins of their father.

**Cast:** *Matt Wolf (Thor), Rick Gomez (Loki), Tara Strong (Sif), Alistair Abell (Fandral), Paul Dobson (Hogun), Brent Chapman (Volstagg), Chris Britton (Odin), Ron Halder (Algrim), Cathy Weseluck (Brunnhilde),* Michael Dobson (Geimarr), John Novak (Thrym), *Ashleigh Ball (Amora),* Brian Drummond (Fenris), Mark Acheson, Mark Gibbon, Jillian Michaels, Ty Olsson, Venus Terzo

**Dir:** Sam Liu; **Writers:** Greg Johnson, Craig Kyle; **Prod:** Gary Hartle; **Music:** Guy Michelmore; **Exec.Prod:** Kevin Feige, Eric S Rollman; **Ed:** George P Rizkillah; **Lionsgate/MLG Productions 7 Inc/Marvel Animation/Marvel Studios**; 77

**Notes:** Completed as far back as 2009, it was decided to wait until **Thor** was released before this one was, in order to capitalise on the publicity.

**Comic Notes:** *Brunnhilde* is a fascinating character, first appearing in *The Avengers #83 (Dec, '70),* created by Roy Thomas and John Buscema.

However, this wasn't actually her – Brunnhilde had formed a partnership with Amora, the Enchantress, which went sour and the Enchantress imprisoned her and impersonated her, adopting the name Valkyrie. Samantha Parrington was transformed by the Enchantress into Brunnhilde's double – *Incredible Hulk #142 (Aug, '71)*, this time Thomas and Herb Trimpe – and she ended up fighting the Hulk. Parrington would later believe she was the real Valkyrie, thanks to Lorelei transforming her, and she would be forced to battle the Defenders. Brunnhilde herself had her spirit enter the body of Barbara Norris, and it was in this body she adopted the Valkyrie mantle (*The Defenders #3 (Dec, '72)*, authored by Steve Englehart and Sal Buscema). Later Brunnhilde escaped her imprisonment and feeling estranged from Asgard, remained with the Defenders (*The Defenders #109 (Jul, '82)*, thanks to Mark Gruenwald, J M DeMatteis and Don Perlin). When Samantha Parrington fought the Defenders, she shook off her control of Lorelei and fought alongside of them, impressing Brunnhilde so much, the goddess elected to leave the Defenders, with Samantha Parrington remaining as Valkyrie. Brunnhilde always claimed to be more than a simple Asgardian, but sadly she was killed by Loki during Ragnarök. Happily, as the other Asgardians were resurrected, so was Brunnhilde.

**Ratings:** *IMDB:* 64%; *Rotten Tomatoes:* 44%

## IRON MAN: RISE OF TECHNOVORE
### *(16/4/2013)*

When Ezekial Stane uses his Technovore armour to attack the launch of Tony Stark's new satellite, Stark himself gets the blame, and becomes the target of SHIELD's manhunt as a result. On the run from his former colleagues, Stark finds himself teaming up with the morally dubious Punisher to find Stane and clear his name.

**Cast:** *Matthew Mercer & Keiji Fujiwara (Tony Stark/Iron Man), Norman Reedus & Tesshō Genda (Frank Castle/The Punisher), Eric Bauza & Miyu Irino (Ezekial Stane/Technovore), Kate Higgins & Megumi Oka (Pepper Potts), James Mathis III & Hiroki Yasumoto (James Rhodes/War Machine), Kari Wahlgren & Junko Minagawa (Maria Hill), Clare Grant & Miyuki Sawashiro (Natasha Romanoff/Black Widow), Troy Baker & Shūhei Sakaguchi (Clint Barton/Hawkeye), Troy Baker & Yasuyuki Kase (JARVIS), Tara Platt & Houko Kuwashima (Sasha Hammer), JB Blanc & Takaya Hashi (Obadiah Stane), John Eric Bentley & Hideaki Tezuka (Nick Fury)*

**Dir:** Hiroshi Hamasaki, Mary Elizabeth McGlynn; **Writers:** Brandon Auman, Kengo Kaji, Steve Kramer; **Prod:** Taro Morishima, Scott Dolph, Megan Thomas Bradner, Harrison Wilcox; **Music:** Tetsuya Takahashi; **Exec.Prod:** Masao Takiyama, Hiroyuki Okada, Alan Fine, Jeph Loeb, Dan Buckley, Simon Phillips; **Ed:** Kashiko Kimura, Mariko Tsukatsune; **Madhouse/Sony Pictures Entertainment (Japan) Inc/Marvel Animation**; 88

**Notes:** A unique anime version of Iron Man, created by a Japanese team, and later released in America with an English speaking cast (hence the dual credits in the cast listing). The series comes from Madhouse who were responsible for the *Marvel Anime* television series.

**Comic Notes:** *Ezekial Stane* – better known as just Zeke – debuted in *The Order #8 (Apr, '08 – Vol 2)* and was created by Matt Fraction and Barry Kitson. Although the movie is a little unclear about his motivation, in the comics he is definitely obsessed with avenging his late father, and does this through cybernetically enhancing a variety of people, including himself. He tends to work behind the scenes, and has teamed up with the Mandarin and the Hellfire Club. Sasha Hammer first appeared in *The Invincible Iron Man #1 (Jul, '08)*, and was created by Fraction and Salvador Larroca. It's not explained in the movie at all, but she is the granddaughter of Justin Hammer, and like the movie she is the girlfriend of Zeke Stane. Her father was the Mandarin, of all people, and along with the rest of her family she has a huge vendetta against Tony Stark and often uses the Detroit Steel armor. Rather creepily, Stane also cybernetically enhanced her, allowing her to fly and create an energy whip.

**Ratings:** *IMDB:* 53%

**Review:** This is *very* anime, ranging from Ezekial Stane and Sasha Hammer's long cryptic conversations, to a tentacled mutant creature at the end, and of course the actual art style. As such, if you don't like anime, you're not going to enjoy this. Having said that, aside from a bizarre plot development (it makes no sense at all that anyone would suspect Stark of attempting to destroy his own project), it's not a terrible film, and Norman Reedus is perfect as the voice of the Punisher.

## IRON MAN & HULK: HEROES UNITED
### *(3/12/2013)*

A Hydra experiment infusing a Stark arc reactor with gamma radiation results in the creation of an insane creature called Zzzax, which immediately begins a reign of destruction. With little choice, Iron Man and Hulk join forces to confront Zzzax, even as Hydra starts to pit its own weapons against them.

**Cast:** *Adrian Pasdar (Tony Stark/Iron Man), Fred Tatasciore (Hulk), Dee Bradley Baker (Zzzax & Dr Cruler), Robin Atkin Downes (Abomination & Dr Fump), David Kaye (JARVIS), Liam O'Brien (Red Skull)*

**Super.Dir:** Leo Riley; **Writers:** Henry Gilroy, Brandon Auman; **Prod:** Ken Duer, Kenneth T Ito; **Music:** Michael McCuistion, Kristopher Carter, Lolita Ritmanis; **Exec.Prod:** Alan Fine, Dan Buckley, Joe Quesada, Jeph Loeb, Stan Lee; **Ed:** Emily Chiu, Jacob Kindberg, Jonathan Polk, Adam Redding; **Marvel Animation/Brian Zoo Studios**; 71

**Notes:** The first animated film using the newer version of the Marvel cartoon universe; as such the cast are from ***Avengers Assemble***.

**Comic Notes:** Zzzax's backstory is obviously not the same in the comics as it is in this movie. Created by Steve Englehart and Herb Trimpe, he first appeared in *The Incredible Hulk #166 (Aug, '73)*. He was created when terrorists destroyed the dynamos at the Con Ed nuclear plant in New York City (which, when you think about it, is probably the least worst outcome of that event). Zzzax has been a comic villain standard, mostly fighting various versions of the Hulk, but also the Avengers, and even Iron Man. Perhaps of most interest is the fact he was once possessed by General "Thunderbolt" Ross, who wanted to destroy the Rick Jones' Hulk.

**Ratings:** *IMDB:* 47%

**Review:** The animation style of this is a little too smooth, similar to ***Spider-Man: The New Animated Series***, which doesn't appeal to me, nor does it suit the characters. Which is a shame, because the story is actually quite good. Pasdar and Tatasciore deliver their usual solid performances, and there's some nice humour in it as well. The fact that, for most of the movie, it's Stark, Hulk and Zzzax, does make it feel cheap, but if you can get past that and the animation, it's quite entertaining.

# AVENGERS CONFIDENTIAL: BLACK WIDOW & PUNISHER
## *(25/3/2014)*

Black Widow arrests and detains the Punisher at SHIELD headquarters, but the terrorist activities of the group Leviathan sees SHIELD technology on the black market. With little choice, Fury decides to team up Black Widow with the Punisher in the hope of them stopping Leviathan.

**Cast:** *Jennifer Carpenter & Miyuki Sawashiro (Natasha Romanoff/Black Widow), Brian Bloom & Tesshō Genda (Frank Castle/The Punisher), Matthew Mercer & Keiji Fujiwara (Tony Stark/Iron Man), Kari Wahlgren & Junko Minagawa (Maria Hill), John Eric Bentley & Hideaki Tezuka (Nick Fury), Grant George & Hiroki Tōchi (Elihas Starr),* Kyle Herbert & Ryūzaburō Ōtomo (Cain), *Eric Bauza & Daisuke Namikawa (Amadeus Cho),* JB Blanc (Orion), *Fred Tatasciore & Yuichi Karasuma (Hulk),* Fred Tatasciore & Hisashi Izumi (Ren), *Matthew Mercer & Shūhei Sakaguchi (Clint Barton/Hawkeye)*

**Dir:** Kenichi Shimizu; **Writers:** Mitsutaka Hirota; **Prod:** Taro Morishima, Scott Dolph, Megan Thomas Bradner, Harrison Wilcox; **Music:** Tetsuya Takahashi; **Exec.Prod:** Masao Takiyama, Hiroyuki Okada, Alan Fine, Jeph Loeb, Dan Buckley, Simon Phillips; **Ed:** Kashiko Kimura, Mariko Tsukatsune; **Madhouse/ Sony Pictures Entertainment (Japan) Inc/Marvel Animation**; 83

**Notes:** Madhouse's second Marvel project. There are uncredited appearances by Thor, War Machine, Captain Marvel and a variety of villains. Impressively the Japanese cast is the same as the previous film, though somewhat less impressively, the American voice cast is not. There are a number of non-speaking cameos – Orion's buyers include Baron Zemo, Griffin, Grim Reaper, Count Nefaria and Graviton, while the Avengers line-up sees Thor, War Machine and Captain Marvel alongside Hawkeye, Iron Man and the Hulk.

**Comic Notes:** Outside of being a genius inventor, there's not an awful lot of similarities between the comic and movie versions of Elihas Starr. He first appeared in *Tales to Astonish #38 (Dec, '62),* and was created by Stan Lee and Jack Kirby. He was originally an atomic scientist who decided to sell secrets, and was ultimately arrested and found guilty of treason. The criminal community arranged his freedom in return for him working for them, and gave him the unfortunate nickname Egghead (and depending on the artist, this can be a literal interpretation). He's worked with a lot of villains, but seems to most often come into conflict with Ant-Man. Amadeus Cho first appeared in *Amazing Fantasy #15 (Jan, '06),* created by Greg Pak and Takeshi Miyazawa. In the comics he is a genius and became close friends with the Hulk and Hercules.

He has been mostly a supporting character during that time, but of note is that since the *All-New All-Different* Marvel comics' event, Cho has become the Hulk. This has not been explained.

**Ratings:** *IMDB:* 58%

**Review:** This one should probably be called Black Widow Vs Punisher seeing as they do seem to spend an awful lot of time fighting each other, for various reasons. Unfortunately it all gets a bit repetitive, and when the Elihas Starr storyline kicks in, the whole thing seems a little ridiculous. Worse than that, around the middle it actually starts to get a bit boring. Add to that some very ordinary voice work (why couldn't they get Norman Reedus back as the Punisher?), and it's not particularly worthy.


## IRON MAN AND CAPTAIN AMERICA: HEROES UNITED
### *(29/7/2014)*

Tony Stark and Steve Rogers find themselves at odds over their approach to situations, but soon the pair find themselves tested when the Taskmaster steals Stark's technology and kidnaps Rogers for the Red Skull's latest scheme. Stark sets out to find his friend and stop the Skull, with only Rogers' approach to a problem available to him.

**Cast:** *Adrian Pasdar (Tony Stark/Iron Man), Roger Craig Smith (Steve Rogers/ Captain America), Fred Tatasciore (Hulk), David Kaye (JARVIS), Clancy Brown (Taskmaster), Liam O'Brien (Red Skull),* Dee Bradley Baker (Dr Cruler), Robin Atkin Downes (Dr Fump)

**Dir:** Leo Riley; **Writers:** Henry Gilroy, Brandon Auman; **Prod:** Ken Duer, Kenneth T Ito; **Exec.Prod:** Alan Fine, Dan Buckley, Joe Quesada, Jeph Loeb, Stan Lee; **Ed:** Adam Redding, Fred Udell; **Marvel Animation**; 71

**Notes:** The second animated film using the newer version of the Marvel cartoon universe. Poor reviews meant this was the last in the series. The film uses music from **Iron Man, The Incredible Hulk, Captain America: The First Avenger, Ultimate Avengers: The Movie** and **Ultimate Avengers 2: Rise of the Panther**.

**Comic Notes:** Tony Masters, the Taskmaster, first appeared in *The Avengers #195 (May, '80)*, created by David Michelinie and George Pérez. His powers are much the same – photographic reflexes – though his backstory is a little different. After his powers were discovered when he copied cowboy television shows, he briefly wanted to be a crime fighter, but decided crime was more profitable, and later decided to start an organisation to train armies for criminals. Over time, he has been used by many villains (though his organisation didn't really take off, he himself was in high demand), but later he was used by SHIELD and given a full pardon when it worked out. He fell in love with his SHIELD handler, Mercedes Merced, and they later married. Of late, however, Taskmaster's allegiances seem to have gone back to the criminal side.

**Ratings:** *IMDB:* 56%

**Review:** There are a few problems with this film, particularly if you have actually watched any of **Avengers Assemble**; the entire storyline surrounding Rogers and Stark's different approaches to fighting was already done in an episode there. Additionally, it again uses the 3D-Wrap process, so the mouth movements of the characters seems stilted. That aside, the voice work is good, and the use of Red Skull and Taskmaster is handled nicely. There are worse ways to pass seventy minutes.

# BIG HERO 6
### *(23/10/2014)*

After visiting his brother at university, Hiro Hamada becomes fascinated by their work and resolves to earn a place at the university, though his devastated when his brother is killed attempting to rescue his mentor in a fire. Hiro accidentally activates his brother's project Baymax, as well as discovering that someone is creating millions of the microbots he himself designed. After being attacked by a man in a Kabuki mask, Hiro gets help from his brother's friends to go after the Kabuki man and find out what really happened to Tadashi Hamada.

**Cast:** *Scott Adsit (Baymax), Ryan Potter (Hiro), Daniel Henney (Tadashi), T J Miller (Fred), Jamie Chung (Go Go), Damon Wayans Jr (Wasabi), Genesis Rodriguez (Honey Lemon),* James Cromwell (Robert Callaghan), Alan Tudyk (Alistair Krei), Maya Rudolph (Cass), Abraham Benrubi (General), Katie Lowes (Abigail)

**Dir:** Don Hall, Chris Williams; **Writers:** Jordan Roberts, Robert L Baird, Daniel

Gerson, Joseph Mateo, Paul Briggs; **Prod:** Roy Conli; **Music:** Henry Jackman; **Exec.Prod:** John Lasseter; **Prod.Des.:** Paul A Felix; **Ed:** Tim Mertens; **Walt Disney Animation Studios/Walt Disney Pictures**; 102

**Notes:** The first proper Disney animation of a Marvel movie, there are a number of references to previous Disney films (**Tangled** and **Wreck-It Ralph** are both notably represented), alongside slight parodies of Marvel films (the post-credits sequence, and the appearance of Stan Lee), the traditional hidden Mickey Mouse silhouettes, and rather obscurely a Dalek from *Doctor Who*. The film won the Academy Award for Best Animated Film at the 87th Academy Awards.

**Stan Spotting:** Fred's Dad in the post-credits sequence – come on, it was in the notes!

**Comic Notes:** There have been two Big Hero 6 mini-series, both of which influence this film. The original team, which first appeared in *Sunfire & Big Hero 6 #1 (Sep, '98)* were created by Steven T Seagle and Duncan Rouleau. They were essentially a Japanese superhero team formed by the government (via a consortium known as the Giri). Silver Samurai was appointed to lead the team (more on him back in **The Wolverine**), and were assisted by Sunfire, Japan's most notable superhero. Both the Samurai and Sunfire's movie rights belong to Fox and so they were never going to be in this film. Making their debut in the same issue, and obviously created by Seagle and Rouleau, were Hiro, Baymax, Fred, Gogo and Honey Lemon. At the age of 13, Hiro Takachiho (not Hamada) was approached by the Silver Samurai to join Big Hero 6, but refused until his mother was kidnapped by the EverWraith, whereupon he agreed to join. Like his movie counterpart, he was extraordinarily brilliant, though his parents were industrialists, and he does not have an older brother (alive or dead). The comic Baymax is radically different to the movie version. Designed by Hiro for a science project, he has a number of external forms which he can morph into, including a dragon and a giant mecha. His basic form looks nothing like the inflatable robot of the movie, and he is not a health care droid. Fredzilla, as he is nicknamed in the comics, actually has the ability to generate a giant lizard creature aura that is solid. Obviously this is referenced by the lizard suit he wears in the movie, but in the comics he is Ainu (an indigenous Japanese ethnicity), and was raised on a SHIELD base. Leiko Tanaka was a young street urchin who became involved with the Yakuza and was arrested and sentenced to five years in prison. Due to her motorcycling skills and good behaviour she was given the chance to get out early by test piloting the GoGo Tomago exo-suit, which she agreed to, and the Giri then recruited her to Big Hero 6. She initially was difficult to work with, but ultimately came to respect her colleagues. Aiko Miyazaki is both very different and very similar to her movie counterpart. Like the movie version, the comic version is both extremely brilliant and very compassionate, though the comic version is more experienced in creating wormholes, one of which is in her purse, and therefore allows her to put virtually anything into it. Extraordinarily beautiful, she was recruited to the Japanese secret service as a research developer, but then lobbied to be a member of Big Hero 6, adopting the codename Honey Lemon. Hiro has quite a crush on her, though GoGo initially can't stand her because of her beauty. Wasabi-no-Ginger first appeared in *Big Hero 6 #1 (Sep, '08)* and was created by Chris Claremont and David Nakayama. Radically different to his movie version, he is Japanese in the comics and is a skilled swordsman, as well as being able to fling knives of Qi energy at his enemies. Tadashi Hamada has never appeared in the mainstream Marvel universe, but first appeared in the comic *Baymax (Aug, '14)* by Haruki Ueno, the official prequel to the movie, where, of course, he was exactly as he appeared in the movie!

**Ratings:** *IMDB:* 79%; *Rotten Tomatoes:* 89%

**Review:** A hugely entertaining children's film that sets out, really, to be something very different to the Manga that inspired it, and as such is a wonderfully fun adventure story. Fans of the comic will obviously be disappointed, but this is definitely not a film aimed at them – or indeed adults in general. The design work is typically Disney in style, which is possibly the only aspect of the film that feels slightly unoriginal, but all other choices give it a uniqueness that makes it very enduring.

## LEGO MARVEL SUPER HEROES: AVENGERS REASSEMBLED
### *(16/11/2015 – Disney XD)*

Whilst setting up for a surprise party, Tony begins to act oddly, and it soon becomes clear he is under the control of Ultron, thanks to Yellowjacket's sabotage. With Ultron planning on taking over the world again, the Avengers launch into action to stop him.

**Cast:** *Laura Bailey (Black Widow), Troy Baker (Hawkeye), Eric Bauza (Iron Spider), Ben Diskin (Spider-Man), Grant George (Ant-Man), JP Karliak (Baron*

Strucker, Vision), Jim Meskimen (Ultron), Bumper Robinson (Falcon), Roger Craig Smith (Captain America), Fred Tatasciore (Hulk), Travis Willingham (Thor, Yellowjacket), Mick Wingert (Iron Man)

**Dir:** Rob Silvestri; **Writer:** Mark Hoffmeier; **Prod:** Leo Martin; **Music:** Asher Lenz & Stephen Skratt; **Exec.Prod:** Jill Wilfert, Jason Cosler, Alan Fine, Dan Buckley, Joe Quesada, Jeph Loeb, Kallan Kagan, Marianne Culbert; **Prod.Des.:** Chi Woo Park; **Ed:** Matt Ahrens; **Marvel Entertainment/Arc Productions**; 22

**Notes:** A short made-for-television film (22 minutes) using the animation style from the Lego Marvel Super Hero games.

**Ratings:** *IMDB:* 69%

**Review:** If you enjoy the Lego Marvel games, you'll enjoy this fun little vignette, which makes the most of the Lego style gags that are included in the games, to build this short story. It's fun, and kids will love it, but it's not anything deep.

## MARVEL SUPER HERO ADVENTURES: FROST FIGHT
### *(11/12/2015)*

Frost Giant Ymir suggests to Loki they should harness the power of Jolnir, using one of the Ancient Caskets. When Thor learns of this plan despite the fact Jolnir is Santa Claus and Tony Stark doesn't believe in him, the Avengers travel to Afalheim to protect him. Meanwhile Rocket Racoon and Groot learn of a bounty for Jolnir and soon they too are after the mythical figure.

**Cast:** *Mick Wingert (Iron Man/Athidel), Matthew Mercer (Captain America/ Gingerbread Man), Travis Willingham (Thor), Fred Tatasciore (Hulk/Ymir), Grey Griffin (Captain Marvel), Troy Baker (Loki), Antony Del Rio (Reptil), Trevor DeVall (Rocket Raccoon/Jarvis/Malitri), Kevin Michael Richardson (Groot), Steve Blum (Santa Claus (Jolnir)/Nick), Jane Singer (Mrs Claus)*

**Dir:** Mitch Schauer; **Writer:** Mark Banker; **Prod:** Kenneth T Ito, Howard Schwartz; **Music:** Kristopher Carter, Lolita Ritmanis, Michael McCuistion; **Exec.Prod:** Alan Fine, Dan Buckley, Joe Quesada, Jeph Loeb, Cort Lane, Stephen Walker; **Marvel Entertainment**; 73

**Notes:** A reasonably length animated movie, this seems to spin off from both **Avengers Assemble** and **Guardians Of The Galaxy**, but as Rocket and Groot are travelling alone as bounty hunters, it looks like it takes place before the movie.

**Ratings:** *IMDB:* 69%

**Review:** This is very kid-friendly, and far more so than even the cartoon series that it spins off from. As such, if you can get past the very saccharine storyline, it's not bad, but I have to admit I couldn't. That said, it's fun when they use Avengers that are relatively unknown such as Captain Marvel and Reptil, and it does make a pleasant change from the cartoon in that regard. Maybe worth watching with kids, but not really worth seeking out.

## HULK: WHERE MONSTERS DWELL
### *(21/10/2016 – Disney XD)*

As Dr Strange battles a number of monstrous creatures, he summons help from Jasper Sitwell's SHIELD Paranormal group as well as the Hulk, but is surprised when, in the middle of battle, Hulk passes out and becomes Banner. Revealing that a child has been lost to Nightmare, Strange tells Banner he thinks that Nightmare has him trapped, but he's not sure if it's Banner who is imprisoned, or the Hulk.

**Cast:** *Fred Tatasciore (Hulk, Countdown), Jesse Burch (Bruce Banner), Liam O'Brien (Dr Stephen Strange), Matthew Waterson (Nightmare), Edward Bosco (Warwolf, Minotaur), Jon Olson (Man-Thing, Rrorgg, Sporr, Zzutak), Michael Robles (Benito Serrano), Mike Vaughn (Jasper Sitwell), Chiari Zanni (Nina Price, Bee Girl), Laura Bailey (Ana), Hope Levy (Gayle), Zach Callison (Eric)*

**Dir:** Mitch Schauer; **Writers:** Marty Isenberg, Dave McDermott; **Prod:** Kenneth T Ito, Howard Schwartz; **Music:** Michael McCuistion, Kristopher Carter, Lolita Ritmanis; **Exec.Prod:** Greg Harman, Cort Lane, Alan Fine, Dan Buckley, Joe Quesada, Jeph Loeb; **Marvel Animation**; 75

**Notes:** A Halloween animated film, which was initially shown at New York ComicCon in 2026.

**Ratings:** *IMDB:* 55%

**Review:** There's quite a cool idea behind this film, representing the difficulties between Banner and the Hulk, and going to some length to explain that Banner seems to be content with the Hulk, whereas the Hulk doesn't feel quite the same. But that's about the depth that the film goes into and as such it feels a little light on. Fun, interesting, but not terribly clever.

## LEGO MARVEL SUPER HEROES – GUARDIANS OF THE GALAXY:

### THE THANOS THREAT
#### *(9/12/2017 – Disney XD)*

Ronan and Nebula seek the Build Stone, something the Guardians currently possess and are in the process of delivering to the Avengers, while avoiding Yondu. When the Guardians are split up, Thanos is finally able to get his hands on the Stone and uses it to build the Big Laser Thingy.

**Cast:** *Willl Friedle (Peter Quill/Star-Lord), Jonathan Adams (Ronan the Accuser), Trevor Devall (Rocket Raccoon), Jennifer Hale (Mantis),* Stan Lee (Ravager Pilot), *Vanessa Marshall (Gamora), Kevin Michael Richardson (Groot), Isaac C Singleton Jr (Thanos), David Sobolov (Drax the Destroyer), Cree Summer (Nebula), James Arnold Taylor (Yondu), Travis Willingham (Thor, Taserface)*

**Dir:** Michael D Black; **Writer:** Mark Hoffmeier; **Prod:** Leslie Barker, Joshua Wexler; **Music:** David Wurst, Eric Wurst; **Exec.Prod:** Jill Wilfert, Jason Cosler, Cort Lane, Robert May, Dan Buckley, Joe Quesada; **Art Dir.:** Frank Louis-Marie; **Ed:** Michael D Black; **Marvel Entertainment/Lego Group/Pure Imagination**; 22

**Ratings:** *IMDB:* 69%

**Review:** More of the same Lego fun as the previous movies. Again, this really depends on whether you're prepared to buy into the style of these Lego movies, but if you do, you'll enjoy this one as much as the others.

## LEGO MARVEL SUPER HEROES – BLACK PANTHER: TROUBLE IN WAKANDA
### *(4/6/2018 – Disney XD)*

Thanos is beaten by the Avengers, and when he recovers, Killmonger and Klaue offer an alliance where they will gain Vibranium to empower Thanos. Shuri alerts T'Challa to the fact that the villains are in Wakanda, and together with Okoye, Black Panther sets out to stop the villains.

**Cast:** *James C Mathis III (T'Challa/Black Panther), Laura Bailey (Black Widow), Yvette Nicole Brown (Okoye), Trevor Devall (Ulysses Klaue), Keston John (Erik Killmonger), Daisy Lightfoot (Shuri), Liam O'Brien (Doctor Strange), Isaac C Singleton (Thanos), Roger Craig Smith (Captain America), Fred Tatasciore (Hulk), Travis Willingham (Thor), Mick Wingert (Iron Man)*

**Dir:** Michael D Black; **Writer:** Mark Hoffmeier; **Prod:** Leslie Barker, Joshua Wexler; **Music:** David Wurst, Eric Wurst; **Exec.Prod:** Jason Cosler, Cort Lane, Robert May, Dan Buckley, Joe Quesada; **Art Dir.:** Frank Louis-Marie; **Ed:** Michael D Black, Alex Verlage; **Marvel Entertainment/Lego Group/Pure Imagination**; 22

**Notes:** Another film – **Lego Spider-Man: Vexed by Venom** is expected to be released in 2019.

**Ratings:** *IMDB:* 63%

**Review:** Much of a muchness as the previous film. You're either into this sort of thing, or it will utterly fail to impress you.

## SPIDER-MAN: INTO THE SPIDER-VERSE
### *(14/12/2018)*

When Miles Morales is bitten by a genetically enhanced spider, he gains special abilities, but is unable to help Spider-Man in his fight against the Kingpin. However, after Peter Parker's death, Miles meets another Peter Parker – from a different dimension. As other dimensions cross over, and more Spider-Men come together, Miles realises he has to take on the Spider-Man mantle in his dimension in order to stop the Kingpin destroying the multiverse.

**Cast:** *Shameik Moore (Miles Morales/Spider-Man), Jakes Johnson (Peter B Parker/Spider-Man), Hailee Steinfeld (Gwen Stacy/Spider-Woman),*

*Mahershala Ali (Aaron Davis/The Prowler), Brian Tyree Henry (Jefferson Davis), Lily Tomlin (May Parker), Luna Lauren Velez (Rio Morales), Zoë Kravitz (Mary Jane Watson), John Mulaney (Peter Porker/Spider-Ham), Kimiko Glenn (Peni Parker/Sp//dr), Nicolas Cage (Peter Parker/Spider-Man Noir), Kathryn Hahn (Liv Octavius/Dr Octopus), Liev Schreiber (Wilson Fisk/The Kingpin), Lake Bell (Vanessa Fisk), Jorma Taccone (Norman Osborn/The Green Goblin), Krondon (Tombstone), Joaquín Cosío (Scorpion), Oscar Isaac (Miguel O'Hara/ Spider-Man 2099)*

**Dir:** Bob Persichetti, Peter Ramsey, Rodney Rothman; **Writers:** Phil Lord, Rodney Rothman; **Prod:** Avi Arad, Amy Pascal, Phil Lord, Christopher Miller, Christina Steinberg; **Music:** Daniel Pemberton; **Exec.Prod:** Will Allegra, Brian Michael Bendis, Stan Lee; **Prod.Des.:** Justin K Thompson; **Ed:** Rob Fisher Jr; **Avi Arad Productions/Lord Miller/Pascal Pictures/Sony Pictures Animation/Colombia Pictures Corporation/Marvel Entertainment**; 117

**Notes:** Chris Pine voices Peter Parker in Miles Morales universe (the character is based on the three live action versions of the character, in an attempt to make the perfect Peter). Jorma Taccone also plays the Peter Parker that Spider-Man 2099 meets, but is credited as "Last dude" to hide the fact. The quote "With great power…" is actually Cliff Robertson, lifted from **Spider-Man**. The film is dedicated to both Steve Ditko and Stan Lee, both of whom passed away during the making of the film. There are a huge number of nods to previous versions of Spider-Mans, including having Miles do the same handshake that Peter does in **Spider-Man: Homecoming**, and some of the moments that Peter recalls are clearly from **Spider-Man 2** and **Spider-Man 3** (though slightly altered). The style of the film is deliberately created to mimic comic books, bringing it a unique visual appearance. The film was nominated for a Golden Globe, a BAFTA and an Academy Award in the category of Best Animated Film, all three of which it actually won.

**Stan Spotting:** Stan actually plays two parts in the film – as the shopkeeper who talks with Miles, and also as J Jonah Jameson in the post-credits sequence. Lee always felt he should have played Jameson, as the character was based loosely on him.

**Comic Notes:** *Miles Morales, Rio Morales, Jefferson Davis & Aaron Davis*: Miles first appeared in *Ultimate Fallout #4 (Aug, '11)*, created by Brian Michael Bendis and Sara Pichelli, as the replacement for Peter Parker in the Ultimate Marvel universe (the Ultimate universe had been conceived as a way to get new readers to Marvel who might be daunted at the enormous history behind the "main" universe), when Peter was killed. The plan was to deliver someone different to Peter – a Brooklyn kid with a black father and Puerto Rican mother. Dr Conrad Markus attempted to recreate Peter Parker's abilities, but the samples were stolen by the Prowler (in this universe, Aaron Davis), along with a spider that would ultimately bite Davis' nephew Miles. Aaron Davis first appeared in *Ultimate Comics: Spider-Man #1 (Nov, '11)*, also created by Bendis and Pichelli (it's worth noting that the Prowler in the mainstream universe is definitely not Aaron Davis). Rio Morales and Jefferson Davis make their first appearances in the same issue. There are a lot of parallels between the movie and the comics – Miles fights Prowler, resulting in his death (though Aaron Davis learns Miles's secret and says they are alike), and Aaron and Jefferson worked together in Turk Barrett's gang, before they were arrested and Jefferson went straight. Later, when Venom attacked, Jefferson was badly hurt, and the fight between Spider-Man and Venom resulted in the death of Rio causing Miles to give up Spider-Man for a year. Jefferson learnt the truth of his son, and blamed him for the deaths of Rio and Aaron. A storyline called *Secret Wars II* resulted in the Ultimate universe ending, and Molecule Man resurrecting both Rio and Aaron, and transferring them and Miles and Jefferson to the mainstream universe before the Ultimate universe collapsed. Here, Miles joined the Avengers and shared the Spider-Man mantle with Peter Parker. After *Civil War II*, he joined the Champions.

*Gwen Stacy:* is an alternative version of the mainstream Gwen, and first appeared in *Edge of Spider-Verse #2 (Sep, '14)*, created by Jason Latour and Robbi Rodriguez. Her background is pretty much what was stated in the movie – she was bitten by the radioactive spider that would have bitten Peter and became Spider-Woman. Peter was killed by the Lizard, and Spider-Woman got the blame for it. Her father, Captain Stacy, discovered her true identity, but maintains his dislike of vigilantes. She has had several versions of her – one who is in a relationship with Miles Morales – and has accrued a variety of nicknames, including Spider-Gwen and Ghost Spider.

*Peni Parker:* is another of the Spiders who debut in *Edge of Spider-Verse - #5 (Dec, '14)* created by Gerard Way and Jake Wyatt. Her father was the pilot of the SP//dr suit, and when he died, she was adopted by her Uncle Ben and Aunt May, and encouraged to carry the project on, despite her young age. She has been forced to fight her friend Addy Brock, who uses the VENOM suit, and been recruited to help multi-verse troubles twice by...

*Peter Porker:* comes from a time when Marvel made comedy spoofs of their material, and first appeared in *Marvel Tails Starring Peter Porker, The Spectacular Spider-Ham (Nov, '83)* created by Tom DeFalco and Mark Armstrong. Peter was actually a spider who was bitten by May Porker after she was irradiated by her atomic hairdryer. This turned him into a pig, who had the abilities of a spider. He became Spider-Ham and fought evil, including villains such as Ducktor Doom, Bull-Frog and the King-Pig. He has become part of mainstream Marvel's adventures, thanks to the *Spider-Verse* series, and currently serves with the Web Warriors – a team of six different Spider-Men, including Spider-Gwen and…

*Spider-Man Noir:* is Peter Parker from the 1930's where he was the apprentice to a reporter, and was bitten by an imported, venomous spider that gave him his abilities (including web generation). He first appeared in *Spider-Man Noir: #1 (Feb, '09)*, part of the "noir" universe, created by David Hine, Fabrice Sapolsky and Carmine di Giandomenico. There have been two series featuring the character, one going up against the crime lord known as the Goblin, the second fighting Doctor Octavius. The *Spider-Verse* series saw him unite with other characters and he currently serves with the Web Warriors, as mentioned above. Though his costume is still black and grey in the comics, it is not assumed his universe is black and white, unlike this movie.

*And:* most of the other characters in this movie appear in earlier Spider-Man films (or Daredevil in one case) and their comic notes can be found earlier in this book.

**Ratings:** *IMDB:* 87%; *Rotten Tomatoes:* 97%; *Metacritic:* 87

**Review:** Genuinely one of the best Spider-Man movies ever made, the idea of leaping through the multiverse and bringing us different versions of Spider-Man is novel (this is the first time it was done), but the decision to concentrate the film on Miles Morales is an excellent one, as it gets to tell the Spider-Man story in a totally new way. There are some brilliant performances, particularly from Moore and Ali, and some amazing use of animation. An absolutely superb film.

## SPIDER-MAN: ACROSS THE SPIDER-VERSE
### *(30/5/2023)*

Trying to balance his life between Spider-Man and passing in school, Miles finds himself at odds with his parents, so that when Gwen Stacy walks back into his life, he immediately decides to follow her. He's surprised to find that she is part of a multi-versal group of Spider-Men, under the leadership of Miguel O'Hara, who work together to stop multi-dimensional problems. Unfortunatlely, Miles' most recent enemy – the Spot – has discovered he has the ability to cross through the multiverse, and decides to use this ability to gain more and more power.

**Cast:** Shameik Moore (Miles Morales/Spider-Man), Hailee Steinfeld (Gwen Stacy/Spider-Woman), Brian Tyree Henry (Jefferson Davis), Luna Lauren Vélez (Rio Morales), Jake Johnson (Peter B Parker/Spider-Man), Jason Schwartzman (Dr Jonathan Ohnn/The Spot), Issa Rae (Jess Drew/Spider-Woman), Karan Soni (Pavitr Prabhakar/Spider-Man India), Daniel Kaluuyu (Hobie Brown/Spider-Punk), Lily Tomlin (May Parker), Oscar Isaac (Miguel O'Hara/Spider-Man 2099), Mahershala Ali (Aaron Davis/The Prowler), Kimiko Glenn (Peni Parker/Sp//dr), Greta Lee (Lyla), Shea Whigham (George Stacy), Jorma Taccone (Adriano Toomes/The Vulture & Peter Parker/Spider-Man 1967), Jharrel Jerome (Miles G Morales/The Prowler)

**Dir:** Bob Persichetti, Peter Ramsey, Rodney Rothman; **Writers:** Phil Lord, Rodney Rothman; **Prod:** Avi Arad, Amy Pascal, Phil Lord, Christopher Miller, Christina Steinberg; **Music:** Daniel Pemberton; **Exec.Prod:** Will Allegra, Brian Michael Bendis, Stan Lee; **Prod.Des.:** Justin K Thompson; **Ed:** Rob Fisher Jr; **Avi Arad Productions/Lord Miller/Pascal Pictures/Sony Pictures Animation/Colombia Pictures Corporation/Marvel Entertainment**; 140

**Notes:** The film uses different animation styles to stress the difference between universes; Earth 65 (Gwen's home universe) has a water colour feel, while Neuva York (Miguel's universe) is based on neo-futurist designs. The punk universe took over two years to complete, and Preston Mutanga, a fourteen year old Canadian, impressed the production team with his Lego version of the first trailer so much so they approached him to make a Lego sequence for the film. The Spot's design evolves through the film, starting as though he is just a rough sketch, complete with blue draft lines. This is the longest US animated film ever made.

Unsurprisingly there are an enormous amount of Spider-people in this movie, starting with the Miles Morales version (Earth 1610), Gwen Stacy (Earth 65), the principal Peter Parker from the previous movie, Jessica Drew, Pavitr Prabhakar, Hobie Brown, Miguel O'Hara, and the 1967 animated Spider-Man. All three movie versions – Tobey Maguire, Andrew Garfield and Tom Holland – appear through achival footage, as does Peter Porker and Peter Parker Noir (John Mulaney and Nicholas Cage). Additional Spider-Men include: Ben Reilly/The Scarlet Spider (Andy Samberg), Margo Kess/Spider-Byte (Amanda Stenberg), Patrick O'Hara/The Web-Slinger (Taran Killam), Lego Spider-Man (Nic Novicki), the spectacular Spider-Man (Josh Keaton using audio from *THE SPECTACULAR SPIDER-MAN*), Insomniac Spider-Man (Yuri Lowenthal, again using audio from the Insomniac video game), Ezekial Sims/Spider-Therapist (Mike Rianda), Malala Windsor/Spider-UK (Sofia Barclay), Charlotte Webber/Sun-Spider (Danielle Perez) and Metro-Spider

(Metro Boomin – a version created specifically for the DJ). Elizabeth Perkins voices a "quippy Spider-Person", and also the brief appearance of Aunt May. There are many other Spider-Men who appear, but those will be covered in the Comic Notes. J K Simmons voices the various versions of J Jonah Jameson, Ayo Edebiri; Nicole Delaney and Antonia Lentini are the voices of Glory, MJ and Betty Brant, Gwen's band members; Melissa Stum is the new voice of Mary Jane Parker; Atsuko Okatsuka voices Yuri Watanbe, a police officer; Peter Sohn is the voice of Ganke Lee; Jack Quaid is the voice of the Peter Parker/Lizard in Gwen's universe, and both Kathryn Hahn and Post Malone appear via audio from the previous film.

Other live-action footage includes Kirsten Dunst, Cliff Robertson, Dennis Leary and Emma Stone using footage from previous Spider-Man films as Mary Jane, Uncle Ben, Captain Stacy and Gwen Stacy, while Alfred Molina's previous dialogue is used to voice a comic version of Dr Octopus. Peggy Lu returns to play Mrs Chen, her character from **VENOM**, and Donald Glover returns as Aaron Davis from **SPIDER-MAN: HOMECOMING** (though this time in Prowler costume). Sony's Spider-man-less Spiderverse is identified as Earth 688, while the MCU is confirmed to be Earth 19999.

Footage is also used from *UNLIMTED SPIDER-MAN*

Other Spider-people include Spider-Knight, Spider-Plushie, Spider-Wizard, cyborg Spider-Woman, two Mariachi Spider-Men, Spider-Man from the *ULTIMATE SPIDER-MAN* animated series, and Spider Canada.

**Comic Notes:** *Miguel O'Hara/Spider-Man*: In 1992, Marvel decided to try a run of "future" titles set in 2099, and introduced a new Spider-Man, one Miguel O'Hara. Miguel is the son of Conchata and her boss, with whom she had an affair, though he was raised by the abusive George O'Hara, and lived with him, his mother and his half-brother Gabe. He attended the Alchemax School for Gifted Youngsters, due to his genius, and then went to work for the company, becoming the head of genetic research. He dated a girl named Xina Kwan, who helped develop the Lyla computer system he uses, but then dated and got engaged to Gabe's ex-girlfriend Dana. Somewhat obsessed with the story of Spider-Man, Miguel attempted to create a serum that would give him those powers. Discovering that the company had addicted him to the drug Rapture, Miguel tried to find a cure, but betrayed by his assistant, Miguel's DNA was rewritten to include spider DNA, giving him similar powers to Spider-Man. He would go onto join forces with Dr Doom, the 2099 versions of the Punisher and the X-Men, and a hero called Ravage to fight a version of Asgard created by the megacorps of Earth. He would go onto become the CEO of Alcehmax, meet Peter Parker, meet the 2211 version of Spider-Man, and get stranded in the present day of Earth. Aside from the movie, Miguel was also instrumental in the unification of may versions of Spider-Man from across the multiverse to stop Superior Spider-Man from destroying the Web of Life; the resolution of which meant that Peter Parker was able to get his body back, having had it stolen by Dr Otto Octavius. He first appeared in *Spider-Man 2099 #1 (Nov, '92)* created by Peter David and Rick Leonardi (actually, technically he first appeared a few months earlier in a preview of this comic in *The Amazing Spider-Man #365*, but let's not quibble.)

*The Spot*: or Dr Jonathan Ohnn, first appeared in *Peter Parker, the Spectacular Spider-Man #97 (Dec, '84)*, created by Al Milgrom and Herb Trimpe, though he wasn't named until the following issue. He was indeed a scientist working for the Kingpin, but his aim was to reproduce the abilities of the superhero Cloak. His partial success led to him having spots on him that worked as portals, but on announcing himself as the Spot, he only made Spider-Man laugh at him. He has since been a member of a number of criminal organisations including MODOK's 11, the Legion of Losers and Hammerhead's army, but rarely acts on his own impetus.

*Ben Reilly*: first appears in *The Amazing Spider-Man #149 (Oct, '75)* created by Gerry Conway. Here he is simply a nameless clone of Peter Parker created by the Jackel. The character was brought back almost twenty years later where it was suggested he was, in fact, the real Peter Parker and what we had seen so far was actually a clone. This was resolved, and the clone adopted the name Ben Reilly – after Ben Parker and May Reilly. When Peter steps down as Spider-Man due to May Parker's stroke, Ben steps into Peter's role, becoming the Scarlet Spider (uniquely, the various Spider-Man comic titles at the time did change their titles to reflect this). He has acted as Spider-Man, and also merged with Carnage. He was killed by the Green Goblin, but a new clone was created, with all of Ben's memories. This version has not only been the Scarlet Spider, but also the Jackel and Chasm. His memory has been tampered with on several occasions, such that his relationship with Peter is not always good, and indeed most recently attempted to bring Peter down, but was locked away in Limbo as a consequence, with Peter hoping that he would regain some form of normality.

*Margo Kess*: hails from the Earth 22191 universe, where the majority of people live via their avatars in cyberspace. In order to stop cybercrime, Margo became the costume vigilante Spider-Byte. When Ghost Spider recruited her into the Spider Army to fight the Inheritors, she was able to project her avatar into the real world. She first appeared in *Vault of Spiders #1 (Oct, '18)* created by Nilah Magruder and Alberto Jiménez Alberquerque.

*Patrick O'Hara*: from Earth 31913, Patrick O'Hara was a gunslinger who was given an elixir by Michael Morbius, which gave both him and his horse Widow, spider powers. He later beat Morbius when the scientist tried to absorb the vital energy of several children. He first appeared in *The Amazing Spider-Man Vol 3 #9 (Nov, '14)*, created by Dan Slott and Giuseppe Carmuncoli.

*And:* There are a LOT of Spider-people who make a cameo appearance, so in an attempt to be thorough...

· Jessica Drew is the Spider-Woman of Earth 616, and first appeared in *Marvel Spotlight #32 (Nov, '76)* created by Archie Goodwin, Sal Buscema, Jim Mooney and Marie Severin. Aside from her cameo appearance, she also serves as the inspiration for the Jess Drew that appears in the movie. Jessica was poisoned by radiation at a young age and her father created a serum from arachnid blood to help cure her. This, along with the High Evolutionary's technology gave her spider powers.

· In *What If? #7 (Nov, '77)*, Earth 78127 came into existence when Flash Thompson was bitten by the radioactive spider instead of Peter. He became Captain Spider as a result, but seemingly died at the hands of the Vulture, where Peter unmasked him to reveal the truth. He was created by Don Glut and Rick Hoberg.

· Malala Windsor is the Earth 835 Spider-UK, a member of the Captain Britain Corps (essentially a multiversal group of Captain Britains). She first appeared in *Daredevils #6 (Jun, '83)* created by Alan Moore and Alan Davis.

· Julia Carpenter was the second Spider-Woman in the prime 616 universe, and debuted in *Marvel Super Heroes Secret Wars #6 (Jun, '84)* created by Jim Shooter and Mike Zeck. She was injected with a spider serum on the orders of Val Cooper, which gave her spider powers, and she donned a suit that was similar to Spider-Man's black suit which he started using at that time. She has an extensive history, and has gone by other names including Arachne and Madame Web.

· Spider-Doppelganger (or just Doppelganger) was a creature with six arms, created by Magus in *Infinity War #1 (Apr, '92)* – created for comics by Jim Starlin and Ron Lim. He was killed, but resurrected by Demogoblin as his pet. He tends to also work for Carnage. He also hails from the prime 616 universe.

· The Scarlet Spider is Kaine Parker, one of many clones of Peter Parker created by the Jackel. He has certain abilities Peter doesn't have, including the ability to mark his victims with an acidic handprint. He first appeared in *Web of Spider-Man #119 (Oct, '94)*.

· Spidercide was a shape-changing creature created by the Jackal who was determined to prove that he was the real Spider-Man. He appears in *The Amazing Spider-Man #399 (Jan, '95)* as Spider-Man, but in *New Warriors #61 (May, '95)* identifies as Spidercide. He was created by J M DeMatteis and Mark Bagley.

· Dr Max Borne is a time spinner from the year 2211, and the universe

Earth 9500. He helped the 2099 Spider-Man defeat his Hobgoblin, though was shocked to discover it was his own daughter. He first appeared in *Spider-Man 2099 Meets Spider-Man #1 (Nov, '95)*, created by Peter David and Mike Wieringo.

· May Parker, nicknamed Mayday, is the daughter of Peter and Mary-Jane in the Earth 982 reality. She developed her father's powers and found herself fighting the grandson of Norman Osborn. She first appeared in *What If...? #105 (Dec, '97)* created by Tom DeFalco, Ron Frenz and Mark Bagley.

· Martha – or Mattie – Franklin is the niece of J Jonah Jameson, and the third Earth 616 Spider-Woman on this list. She gained her powers through an arcane ritual and took on the Spider-Woman mantle with Jessica Drew's permission, developing a serious crush on Spider-Man in the process. She died saving the life of Cindy Moon – the superhero known as Silk. She first appeared in *Spectacular Spider-Man #263 (Dec, '98)* created by John Byrne and Rafael Kayanan.

· On Earth 9997, Peter Parker left his superhero life when Norman Osborne revealed his secret identity. He would go on to join the NYPD at Luke Cage's request and on occasion act as Spider-Cop. He first appeared in *Earth X #0 (Jan, '99)* created by Jim Krueger, Alex Ross and John Paul Leon.

· Earth 2301 is known as the Mangaverse, where Peter Parker is part of a ninja clan, the Spider-Clan. He gained some of his powers thanks to being possessed by a member of the Shadow Clan – Venom. He debuted in *Marvel Mangaverse: Spider-Man #1 (Jan, '02)* created by Ben Dunn and Tommy Ohtsuka.

· Peter Benjamin Parker of Earth 312500 was a darker Spider-Man, the result of having killed Kraven the Hunter in revenge. He was ultimately killed by the NYPD. He first appeared in *The Amazing Spider-Man #58 (Sep, '03)* created by J Michael Straczynski and John Romita Jr.

· Aña Corazón is a teenage gymnast first appearing in *Amazing Fantasy Vol 2 #1 (Jun, '04)*, created by Joe Quesada, Fiona Avery and Mark Brooks, though she doesn't appear as Araña until #5. Miguel Legar, the sorcerer of the Spider Society, gave her some of his power to become the Society's new Hunter. She is from Earth 616, though has helped a version of herself going by Spider-Girl in another universe, and is regarded by that Aña as a sister.

· The Pravitr Prabhakar Spider-Man of Mumbattan is based on the Earth 50101 character of the same name, who also makes a brief cameo. First appearing in *Spider-Man: India #1 (Nov, '04)*, and created by Jeevan Kang, Suresh Seetharaman and Sharad Devarajan, this version is a poor boy in Mumbai who gets powers from a local yogi.

· Bruce Banner, as the Hulk, was bitten by a radioactive spider and became the Spider-Man of Earth 200527. He first appeared in *Wha...Huh? #1*

*(Aug, '05)* created by Mark Waid, Tom Peyer and Jim Mahfood.

· In the Earth 1610 universe, a clone of Peter was created that had six arms, though unlike Doppelganger, would wear the black Spider-Man suit. Named Tarantula, this version was obsessed with protecting Mary-Jane, but ostensibly killed by Dr Octopus. He first appeared in *Ultimate Spider-Man #100 (Nov, '06)*, created by Brian Michael Bendis and Mark Bagley.

· In a fairy tale land, Peter Parker constructed a suit of armor to be the Prince of Arachne and win Princess Gwendolyn's hand in marriage. Sadly it was ill-fated as his adopted father, Norman, worked against him. He first appeared in *Spider-Man: Fairy Tales #4 (Sep, '07)*, created by C B Cebulski and Nick Dragotta.

· Ashley Barton, the daughter of Tonya Parker and Hawkeye, became Spider-Woman in order to stop the Kingpin, before taking over his territory. During the battle against Karn, her universe Earth 807128 was destroyed, but reconstituted as Earth 21923. Spider-Kingpin, as she is sometimes known, first appeared in *Wolverine Vol 3 #67 (Jul, '08)* created by Mark Millar and Steve McNiven.

· Spider-Monkey was the Peter Parker of Earth 8101, a universe where primates still ruled. A harsh reality, he helped the Ape-vengers beat Dr Ooktavius to death after the villain had been caught. He first appeared in *The Amazing Spider-Man Family #1 (Aug, '08)* created by Karl Kesel and Ramon Bachs.

· Spider-Cat appeared in *Spider-Island: I Love New York City #1 (Sep, '11)* created by Skottie Young. Having gained his powers from a Spider Totem, he fought a pigeon called Venom in the Earth 999 reality. He was killed by the Inheritors.

· Lady Spider is Maybelle Reilly was taught a valuable lesson when bitten by a caged spider, and so built a suit with four mechanical arms. She hails from Earth 803 and first appeared in *Spider-Verse # (Nov, '14)* created by Robbie Thompson and Denis Medri.

· Spider-Punk is based on the Spider-Punk of Earth 138 (also named Hobart "Hobie" Brown), and is quite similar to the movie version. He was bitten by a spider irradiated from nuclear waste. He first appeared in *The Amazing Spider-Man Vol 3 #10 (Nov, '14)*, created by Dan Slott and Olivier Coipel.

· A werewolf with spider powers, Spider-Wolf was the Spider-Man of Earth 13989. He was also created by Slott and Coipel and first appeared in *The Amazing Spider-Man Vol 3 #11 (Dec, '14)*.

· Peter Parkedcar hails from Earth 53931, a universe of sentient vehicles, where he is the spectacular Spider-Mobile. He first appeared in *The Amazing Spider-Man Vol 3 #12 (Jan, '15)* and was created by Dan Slott and Giuseppe Camuncoli.

Mary-Jane Watson-Parker and Anna-May "Annie" Parker are from Earth 18119, and first appeared in *Amazing Spider-Man: Renew Your Vows #1 (Jun, '15)*. Annie discovered she had powers and then joined with Power Pack to become Spiderling. Concerned by this, Peter Parker created a suit that drew on and mimicked his powers, so that Mary-Jane could keep an eye on Annie as Spinneret. Annie accidentally unleashed the god Shathra, becoming its thrall, but Mary-Jane – now using the Venom symbiote – was able to help stop the god and free her daughter. They were created by Dan Slott and Adam Kubert.

Charlotte Webber, the Sun-Spider, first appears in *Spider-Verse Vol 3 #2 (Dec, '19)* created by Dayn Broder. She suffers from Ehlers-Danlos syndrome, which means she can't use her legs, but her powers compensate for this and she carries a wheelchair on her back. Her universe is designated Earth 20023.

Pter Ptarker, the amazing Spider-Rex, first appeared in *Edge of Spider-Verse Vol 2 #1 (Aug, '22)* created by Karla Pacheco and Pere Pérez. Formerly a Pteranodon, Pter and the T-Rex Norrannosaurman were hit by a meteorite containing alien spiders, which caused them to switch bodies, gave them amazing costumes and also spider-powers. Pter was able to fight Norrannosaurman, but in the process two innocent dinosaurs died and Pter realised he had to use his powers carefully.

Spider Samurai was one of Shathra's enforcers in *Spider-Man Vol 4 #6 (May, '23)* created by Dan Slott and Mark Bagley.

There are also a lot of Spider-Man costumes that aren't really multiversal alternates, but seem to have been included for being unique.

"Bombastic Bag Man" is the Fantastic Four suit with the paper bag that Peter wore in *The Amazing Spider-Man #258 (Nov, '84)* after losing his old suit and gaining the black one. Due to legal reasons, in the movie it doesn't have the 4 logo.

Spider-Armor Mk 1 made Peter invincible and appeared in *Web of Spider-Man #100 (May, '93)*.

Electro-proof Suit was made to battle Elektro in *The Amazing Spider-Man #425 (Aug, '97)*.

The Iron Spider armored suit was built by Tony Stark to help Peter during the Civil War, which pitted heroes against each other. It first appeared in *The Amazing Spider-Man #529 (Apr, '06)*.

The Stealth suit was created to battle Hobgoblin and his sonic screams in *The Amazing Spider-Man #650 (Feb, '11)*.

Spider-Armor Mk 2 (the update on the original) was used in *The Amazing Spider-Man #656 (Mar, '11)*.

Spider-Armor Mk 3, which was built if Peter needed to go up against the Sinister Six, was used in *The Amazing Spider-Man #682 (Mar, '12)*.

·       The Superior suit was used in *The Amazing Spider-Man #700 (Feb, '13)* though it was actually Otto Octavius in Peter Parker's body at the time.

·       Future Foundation Suit was given to Peter when he joined the organisation in *FF #1 (May, '11)*. Though, possibly, this might actually be the Life Model Decoy of Peter who appeared in *Murderworld: Spider-Man #1 (Dec, '22)*.

There are, allegedly, 240 different versions of Spider-Man in this film. This book concedes it may not have listed them all. But seriously…

**Ratings:** *IMDB:* 89%; *Rotten Tomatoes:* 96%; *Metacritic:* 86

**Review:** Whoah! This is a sequel and a half. There's so much crammed in here for Spider-Man fans that it's easy to miss a lot of it, but that doesn't matter, because the story is a clever idea and is a fantastic way of showing how different Miles Morales is to Peter Parker. By the time the movie comes to an end, it's easy to forget this is actually a two-part story, so the cliffhanger works really well, and is a very brilliant way to show off other aspects of the Multiverse. A fantastic sequel.

# IV. ANIMATED TELEVISION SERIES

## THE MARVEL SUPER HEROES
### (September – December, 1966)
The heroes of Marvel comics battle against the villains that attack New York.

**Regular Cast:** *Sandy Becker (Captain America), Max Ferguson (Hulk), John Vernon (Iron Man/Namor), Chris Wiggins (Thor)*

**Episodes:**

*1.1 Captain America: Episode 1*
*1.2 Captain America: Episode 2*
*1.3 Captain America: Episode 3*
*1.4 Captain America: Episode 4*
*1.5 Captain America: Episode 5*
*1.6 Captain America: Episode 6*
*1.7 Captain America: Episode 7*
*1.8 Captain America: Episode 8*
*1.9 Captain America: Episode 9*
*1.10 Captain America: Episode 10*
*1.11 Captain America: Episode 11*
*1.12 Captain America: Episode 12*
*1.13 Captain America: Episode 13*
*1.14 The Incredible Hulk: Episode 1*
*1.15 The Incredible Hulk: Episode 2*
*1.16 The Incredible Hulk: Episode 3*
*1.17 The Incredible Hulk: Episode 4*
*1.18 The Incredible Hulk: Episode 5*
*1.19 The Incredible Hulk: Episode 6*
*1.20 The Incredible Hulk: Episode 7*
*1.21 The Incredible Hulk: Episode 8*
*1.22 The Incredible Hulk: Episode 9*
*1.23 The Incredible Hulk: Episode 10*
*1.24 The Incredible Hulk: Episode 11*
*1.25 The Incredible Hulk: Episode 12*
*1.26 The Incredible Hulk: Episode 13*
*1.27 The Invincible Iron Man: Episode 1*
*1.28 The Invincible Iron Man: Episode 2*
*1.29 The Invincible Iron Man: Episode 3*
*1.30 The Invincible Iron Man: Episode 4*
*1.31 The Invincible Iron Man: Episode 5*
*1.32 The Invincible Iron Man: Episode 6*
*1.33 The Invincible Iron Man: Episode 7*
*1.34 The Invincible Iron Man: Episode 8*
*1.35 The Invincible Iron Man: Episode 9*
*1.36 The Invincible Iron Man: Episode 10*
*1.37 The Invincible Iron Man: Episode 11*
*1.38 The Invincible Iron Man: Episode 12*
*1.39 The Invincible Iron Man: Episode 13*
*1.40 The Might Thor: Episode 1*
*1.41 The Might Thor: Episode 2*
*1.42 The Might Thor: Episode 3*
*1.43 The Might Thor: Episode 4*
*1.44 The Might Thor: Episode 5*

1.45 The Might Thor: Episode 6
1.46 The Might Thor: Episode 7
1.47 The Might Thor: Episode 8
1.48 The Might Thor: Episode 9
1.49 The Might Thor: Episode 10
1.50 The Might Thor: Episode 11
1.51 The Might Thor: Episode 12
1.52 The Might Thor: Episode 13
1.53 Prince Namor, The Sub Mariner: Episode 1
1.54 Prince Namor, The Sub Mariner: Episode 2
1.55 Prince Namor, The Sub Mariner: Episode 3
1.57 Prince Namor, The Sub Mariner: Episode 5
1.58 Prince Namor, The Sub Mariner: Episode 6
1.59 Prince Namor, The Sub Mariner: Episode 7
1.60 Prince Namor, The Sub Mariner: Episode 8
1.61 Prince Namor, The Sub Mariner: Episode 9
1.62 Prince Namor, The Sub Mariner: Episode 10
1.63 Prince Namor, The Sub Mariner: Episode 11
1.64 Prince Namor, The Sub Mariner: Episode 12
1.65 Prince Namor, The Sub Mariner: Episode 13

**Notable Guest Cast:** *Bernard Cowan (Odin), Len Carlson (Quicksilver), Henry Ramer (Crimson Dynamo)*

**Prod:** Steve Krantz; **Exec.Prod:** Robert L Lawrence; **Ed:** Walter B Corso, Hank Gotzenberg, George Mahana; **Grant-Ray Lawrence Animation/Marvel Comics Group**; c 16

**Notes:** The first Marvel cartoon was a series of 65 episodes split into three segments. It was essentially frames from comic books and manipulated (a process called xerography). The art, therefore, was done by the likes of Jack Kirby, Steve Ditko, etc. A large number of Marvel superheroes would appear in the series though.

**Ratings:** *IMDB:* 73%

**Review:** The animation is extremely basic, but the stories aren't that bad, and make good use of the characters from the Marvel universe. On occasion the voice actors seems a little miscast, but surprisingly, most of them work quite well. Given its time, this is a great start to the era of Marvel cartoons.

# FANTASTIC FOUR

## September, 1967 – September, 1968 /ABC)

On a mission in space, Reed Richards, his girlfriend Sue Storm, her brother Johnny and Reed's best friend Ben Grimm, fly through cosmic rays which give them astonishing powers. Once on Earth the four do battle against threats the world over.

**Regular Cast:** *Paul Frees (The Thing), Gerald Mohr (Mr Fantastic), Jack DeLeon (The Human Torch), Jo Ann Pflug (The Invisible Girl)*

**Episodes:**

*1.1 Menace of the Mole Men*
*1.2 Diablo*
*1.3 The Way It All Began*
*1.4 Invasion of the Super-Skrulls*
*1.5 Klaws*
*1.6 The Red Ghost*
*1.7 Prisoners of Planet X*
*1.8 It Started On Yancy Street*
*1.9 Three Predictions of Dr Doom*

*1.10 Behold a Distant Star*
*1.11 Demon in the Deep*
*1.12 Danger in the Depths*
*1.13 Return of the Mole Man*
*1.14 Rama-Tut*
*1.15 Galactus*
*1.16 The Micro World of Dr Doom*
*1.17 Blastaar, The Living Bomb-Burst*
*1.18 The Mysterious Molecule Man*
*1.19 The Terrible Tribunal*
*1.20 The Deadly Director*

**Notable Guest Cast:** *Joseph Sirola (Dr Doom), Marvin Miller (Super Skrull),* Don Messick (Kurrgo), Charles Spidar (Citizen), *Vic Perrin (Silver Surfer), Ted Cassidy (Galactus), Henry Corden (Attuma), Janet Waldo (Lady Dorma), Regis Cordic (Diablo), Frank Gerstle (Blastaar), Mike Road (Prince Triton), Hal Smith (Klaw)*

**Dir/Prod:** William Hanna, Joseph Barbera; **Writers:** Phil Hahn, Jack Hanrahan; **Ed:** Earl Bennett, Milton Krear, Skip Lusk; **Hanna-Barbera Productions/Marvel Enterprises**; 22

**Notes:** The first of the Hanna-Barbera/Marvel programs. Despite a positive reaction to the program, it was cancelled due to the public pressure about violence on television. Like all television shows from this period, it has over time changed hands as various companies have bought, sold and merged with each other. Strangely, while nearly all of Marvel's media projects have been acquired by Disney, this one has not; instead it is owned by Time Warner, the company that, ironically, owns DC.

**Ratings:** *IMDB:* 69%

**Review:** This has to be viewed within context, and based on that a certain number of things, such as the very sixties attitudes towards sexism and racism, are bound to be present. Equally, the characters themselves are very sixties – Reed smokes a pipe, for example! The animation itself is not of a great quality (indeed this series starts a trend of cartoons being unable to do a decent animation of the Thing), and the endings tend to be a little twee, but in spite of all of that, this isn't a bad program, and does a fairly decent job of representing the source material. Certainly better than its successor.

### SPIDER-MAN
#### (September, 1967 – June, 1970/ABC)

Peter Parker, freelance photographer for the Daily Bugle, balances his work life with his secret life as Spider-Man, battling his editor to curb the anti-Spider-Man stories, and battling the super powered menaces that threaten New York.

**Regular Cast:** *Paul Soles (Spider-Man), Bernard Cowan (Dr Doom), Paul Kilgram (J Jonah Jameson), Peg Dixon (May Parker)*

**Episodes:**

*1.1 The Power of Dr Octopus/Sub-Zero for Spidey*
*1.2 Where Crawls the Lizard/Electro, the Human Lightning Bolt*
*1.3 The Menace of Mysterio*
*1.4 The Sky Is Falling/Captured by J Jonah Jameson*
*1.5 Never Step On a Scorpion/Sands of Crime*
*1.6 Diet of Destruction/The Witching Hour*
*1.7 Kilowatt Kaper/The Peril of Parafino*
*1.8 Horn of the Rhino*
*1.9 The One-Eyed Idol/Fifth Avenue Phantom*
*1.10 The Revenge of Dr Magneto/The Sinister Prime Minister*
*1.11 The Night of the Villains/Here Comes Trubble*
*1.12 Spider-Man Meets Dr Noah Boddy/The Fantastic Fakir*
*1.13 Return of the Flying Dutchman/Farewell Performance*
*1.14 The Golden Rhino/Blueprint for Crime*
*1.15 The Spider and the Fly/The Slippery Dr Von Schlick*
*1.16 The Vulture's Prey/The Dark Terrors*

**Notable Guest Cast:** *Vern Chapman (Doc Ock), Gillie Fenwick (Lizard), Tom Harvey (Elektro), Chris Wiggins (Mysterio), Carl Banas (Scorpion), Len*

*Carlson (Green Goblin), Ed McNamara (Rhino)*

**Dir:** Grant Simmons [1.1-1.20], Clyde Geronimi [1.1-1.20], Sid Marcus [1.1-1.20], June Patterson [1.1-1.20], Ralph Bakshi [2.1-3.13]; **Prod:** Ray Patterson [1.1-1.20]; **Writers:** Bill Danch, Al Bertino, Dick Robbins, Dick Cassarino, Phil Babet; **Music:** Ray Ellis; **Ed:** Bryce Corso [1.1-1.20], Howard Kaiser [2.1-3.13], George Copeland [2.1-3.13], Richard Calamari [3.1-3.13]; **Exec.Prod:** Ralph Bakshi [2.1-3.13]; **Grant-Ray Lawrence Animation** [1.1-1.20]/**Krantz, Animation Inc** [2.1-3.13]/**Marvel Comics Group**; 25

**Notes:** ABC originally poured a lot of money into this series, but Grant-Ray Lawrence Animation wasn't able to meet the requirements after the first season, and so Krantz Animation took over. A third season was necessary to make the minimum number of episodes for syndication, but by then the money was falling short and some episodes (like "Rhino") simply recycled footage from previous seasons, while other episodes (like "Revolt In The Fifth Dimension") actually took an episode from another animated series (*Rocket Robin Hood*) and simply stuck Spider-Man over the hero. The Krantz seasons tended to avoid characters from the comics, and went for generic monsters instead. The theme for this series is extremely well known, and incorporated into the Sam Raimi **Spider-Man** films.

**Ratings:** *IMDB:* 75%

**Review:** Not a bad little cartoon, particularly when viewed in context. The animation is probably better in the first season, but it's crude by today's standards regardless, and as such is largely forgivable. Whilst it sometimes lacks a sense of humour, it's a worthwhile series, and quite enjoyable.

## THE FANTASTIC FOUR
### (September – December, 1978/NBC)

On a mission in space, Reed Richards, his girlfriend Sue Storm and his best friend Ben Grimm, fly through cosmic rays which give them astonishing powers. Once on Earth, and accompanied by a robot named HERBIE, the four do battle against threats the world over. Again.

**Regular Cast:** *Mike Road (Reed Richards), Ginny Tyler (Sue Storm), Ted Cassidy (Ben Grimm), Frank Welker (HERBIE), John Stephenson (Dr Doom),*

*Gene Moss (Trapster),* Don Messick (J J Colossal), Hal Smith, Nancy Wible, Joan Gerber, Vic Perrin, Dick Tufeld (Narrator)

**Episodes:**

*1.1 A Monster Among Us*
*1.2 The Menace of Magneto*
*1.3 The Phantom of Film City*
*1.4 Medusa and the Inhumans*
*1.5 The Diamond of Doom*
*1.6 The Mole Man*
*1.7 The Olympics of Space*
*1.8 The Fantastic Four Meet Doctor Doom*
*1.9 The Frightful Four*
*1.10 Calamity on the Campus*
*1.11 The Impossible Man*
*1.12 The Final Victory of Doctor Doom*
*1.13 Blastaar, the Living Bomb Burst*

**Writers:** Stan Lee, Roy Thomas, Bob Johnson, Christy Marx, Bob Stitzel; **Prod:** David H DePatie, Friz Freleng; **Music:** Dean Elliott; **Exec.Prod:** Lee Gunther; **Ed:** Richard Corwin, David H DePatie Jr, Richard S Gannon; **DePatie-Freleng Enterprises/Marvel Comics Animation**; 22

**Notes:** Due to rights surrounding the possibility of a Human Torch television series, the Torch was dropped from this cartoon, and replaced by the robot HERBIE, who would go onto appear in the comics. This is the start of the DePatie-Freleng/Marvel partnership that would produce nearly all animated Marvel until 1989. The narrator, despite the onscreen title, refers to the series as *The New Fantastic Four*.

**Ratings:** *IMDB:* 60%

**Review:** If the inclusion of HERBIE wasn't going to make this series painfully childish, the absurd plots most certainly do (and bizarrely from Stan Lee of all people). Magneto winning the leadership of the Fantastic Four? Dr Doom sending the men back in time to collect Blackbeard's treasure? Then there's the Impossible Man, played straight; and they *still* haven't learnt how to draw the Thing! Very silly.

## FRED AND BARNEY MEET THE THING
### (September – December, 1979/NBC)

Benjy Grimm uses his Thing Rings to stop the mischievous enterprises of the Yancy Street Gang, the bumbling misadventures of snobby Ronald Radford and the occasional mad scientist.

**Regular Cast:** *Wayne Morton (Benjy Grimm,* Stretch), *Joe Baker (The Thing),* Noelle North (Kelly), Marilyn Schreffler (Betty, Miss Twilly), John Erwin (Ronald Radford), Art Metrano (Spike), Michael Sheehan (Turkey), John Stephenson (Dr Harkness)

**Episodes:**

*1.1 The Picnic Panic/Bigfoot Meets the Thing*
*1.2 Junkyard Hijinks/Gone Away Gulch*
*1.3 Circus Stampede/The Thing and the Queen*
*1.4 Canival Caper/The Thing Blanks Out*
*1.5 The Thing Meets the Clunk/Beach Party Crashers*
*1.6 Decepto the Great/The Thing's the Play*
*1.7 Double Trouble for the Thing/To Thing or Not to Thing*
*1.8 The Big Bike Race/The Thing and the Treasure Hunt*
*1.9 Out to Launch/The Day the Ring Didn't Do a Thing*
*1.10 A Hot Air Affair at the Fair/The Thing Goes to the Dogs*
*1.11 The Thing Goes Camping/Dude Ranch Rodeo*
*1.12 Photo Finish/Lights, Action, Thing!*
*1.13 The Thing and The Captain's Ghost/The Thing and the Absent-Minded Inventor*

**Prod:** Alex Lovy; **Exec.Prod:** Joseph Barbera, William Hanna; **Hanna-Barbera Productions**; 60

**Notes:** Contrary to what the title might suggest, this was not a crossover between *The Flintstones* and *Fantastic Four.* In fact, *The Thing* segments were about 12 minutes long, with two per episode, combined with an episode of either *The New Fred and Barney Show*, or a genuine new thirty minute episode featuring the Flintstones. It's not clear how Hanna-Barbera had the rights to the Thing, given that DePatie-Freleng clearly had them a year earlier. Perhaps unsurprisingly the Thing's backstory is completely changed, but oddly enough the name "Yancy Street Gang" does come from the comics; though Spike, Stretch and Turk were never members of it. The following year, the series was repackaged with episodes of *The New Shmoo* as *Fred and Barney Meet the Shmoo*. Little bumper segments between episodes did show the characters meeting briefly. John Erwin would go onto play Prince Adam and He-Man in *He-Man and the Masters of the Universe*. Somewhat strangely, the comic *Marvel NOW! Point One #1 (Dec, '12)* makes reference to this series when Ant-Man gives the pop star Darla Deering a set of rings that when she combines them forms an exo-skeleton around her to replicate the Thing. She even goes so far as to use the television phrase "Thing rings, do your thing!"

**Ratings:** *IMDB:* 65%

**Review:** There's no mistaking this for anything other than a Hanna-Barbera cartoon, which means it's a sort of bizarre combination between Scooby-Doo and Archie, featuring the Thing. In as of itself, that's not a bad thing, but as an adaption of The Thing it leaves a lot to be desired – and the Thing design is predictably awful. As a mindless, entertaining 1970's cartoon, it does what it's supposed to and works well enough.

## SPIDER-WOMAN
### (September, 1979 – January, 1980/ABC)

Bitten by a poisonous spider as a child, Jessica Drew was given an experimental spider serum to cure her, which had the side effect of giving her amazing powers. Now working for Justice Magazine, she fights crime as Spider-Woman to protect the city and her nephew, Billy.

**Regular Cast:** *Joan Van Ark (Jessica Drew/Spider-Woman),* Bryan Scott (Billy), Larry Carroll (Detective Miller), Bruce Miller (Jeff Hunt), Lou Krugman (Chief Cooper), Vic Perrin, Tony Young, John Milford, John H Mayer, Ilene Latter, Karen McMachon, Dick Tufeld (Narrator)

**Episodes:**

| | |
|---|---|
| *1.1 Pyramids of Terror* | *1.8 Games of Doom* |
| *1.2 Realm of Darkness* | *1.9 Shuttle to Disaster* |
| *1.3 The Amazon Adventure* | *1.10 Dracula's Revenge* |
| *1.4 The Ghost Vikings* | *1.11 The Spider-Woman and the Fly* |
| *1.5 The Kingpin Strikes Again* | *1.12 Invasion of the Black Hole* |
| *1.6 The Lost Continent* | *1.13 The Great Magini* |
| *1.7 The Kongo Spider* | *1.14 A Crime in Time* |

*1.15 Return of the Spider-Queen*
*1.16 The Deadly Dream*

**Notable Guest Cast:** *Paul Soles (Spider-Man)*

**Dir:** Bob Richardson; **Writers:** Jeffrey Scott, Tom Swale; **Prod:** Lee Gunther; **Music:** Eric Rogers; **Exec.Prod:** David H DePatie, Friz Freleng; **Ed:** Richard S Gannon; **DePatie-Freleng Enterprises/Marvel Comics Animation**; 30

**Notes:** Not particularly faithful to the comic that spawned it, it's not entirely clear why the companies opted for Spider-Woman rather than Spider-Man, though it probably had something to do with the fact the live-action series was still going. For 1.1 and 1.7, Paul Soles reprised the voice of Spider-Man.

**Ratings:** *IMDB:* 61%

**Review:** A fairly standard run-around, with the writers obviously a bit nervous that without Spider-Man in the first episode, the series wouldn't hit. For the most part it avoids stupid clichés, but inevitably sexism still creeps in, despite this being about Spider-Woman. Marks for at least choosing a unique character to make the series around, but none for originality.

## *SPIDER-MAN*
### *(September, 1981 – March, 1982)*
While Peter Parker tries to balance his real life with his life of crime fighting as Spider-Man, big problems arise in Latveria where Dr Doom's dictatorship is threatened by the rebels; rebels who request help from Spider-Man.

**Regular Cast:** *Ted Schwartz (Peter Parker/Spider-Man), William Woodson (J Jonah Jameson), Linda Gary (Aunt May Parker), Mona Marshall (Betty Brant), Ralph James (Dr Doom)*

**Episodes:**

*1.1 Bubble, Bubble, Oil and Trouble*
*1.2 Dr Doom, Master of the World*
*1.3 Lizards, Lizards, Everywhere*
*1.4 Curiosity Killed the Spider-Man*
*1.5 The Sandman Is Coming*
*1.6 When Magneto Speaks...People Listen*
*1.7 The Pied Piper of New York Town*
*1.8 The Doctor Prescribes Doom*

**Notable Guest Cast:** *Stan Jones (The Kingpin), Peter Cullen (Red Skull), Michael Rye (Magneto), George DiCenzo (Captain America), Wally Burr (The Sandman), Corey Burton (The Lizard), Morgan Lofting (Black Cat), Dennis Marks (Green Goblin), Robert Ridgely (Kazar), Neil Ross (Norman Osborn), Michael Sheehan (Mortimer), Don Messick (Vulture), Paul Winchell (Silvermane)*

**Dir:** Don Jurwich; **Prod:** Art Vitello; **Music:** Johnny Douglas; **Exec.Prod:** David H DePatie, Lee Gunther; **Marvel Productions**; 22

**Notes:** Like *Spider-Woman* this series was syndicated. Peter Cullen would go onto find fame as Optimus Prime in *The Transformers* cartoon and later, **The Transformers** movie series. The artwork for the series is based on John Romita Jr's work in the comics. DePatie-Freleng Enterprises was renamed Marvel Productions at this time.

**Ratings:** *IMDB:* 70%

**Review:** A step up from *Spider-Woman*, certainly helped by the overall story arc, and generally better writing for the episodes as a whole. The series is surprisingly faithful to the comic, though there are some interesting (though not ridiculous) changes; the final episode, for instance, sees Peter falling in love with Medusa, of all people. Worth watching.

## SPIDER-MAN AND HIS AMAZING FRIENDS
### (September, 1981 – September, 1983/NBC)

Spider-Man joins forces with Iceman and Firestar to protect New York from the sinister threats that menace it.

**Regular Cast:** *Dan Gilvezan (Peter Parker/Spider-Man), Kathy Garver (Anjelica Jones/Firestar), Frank Welker (Bobby Drake/Iceman), June Foray (Aunt May),* Stan Lee (Narrator), Dick Tufeld, William Marshall, Ron Feinberg (Announcers)

**Episodes:**

*1.1 Triumph of the Green Goblin*
*1.2 The Crime of All Centuries*
*1.3 The Fantastic Mr Frump!*
*1.4 Sunfire*
*1.5 Swarm*
*1.6 7 Little Superheroes*
*1.7 Videoman*
*1.8 The Prison Plot*
*1.9 Spidey Goes Hollywood*
*1.10 The Vengeance of Loki!*
*1.11 Knights and Demons*
*1.12 Pawns of the Kingpin*
*1.13 The Quest of the Red Skull*

*2.1 The Origin of the Iceman*
*2.2 Along Came Spidey*
*2.3 A Fire-Star Is Born*
*3.1 Spider-Man Unmasked*
*3.2 The Bride of Dracula!*
*3.3 The Education of a Superhero*
*3.4 Attack of the Arachnoid*
*3.5 The Origin of the Spider-Friends*
*3.6 Spidey Meets the Girl from Tomorrow*
*3.7 The X-Men Adventure*
*3.8 Mission: Save the GuardStar*

**Notable Guest Cast:** *John Stephenson (Colossus), Neil Ross (Wolverine), George DiCenzo (Captain America), Anne Lockhart (Storm), William Woodson (J Jonah Jameson), Stan Jones (Professor Xavier), Walker Edmiston (Kingpin),* Michael Ansara (Hiawatha Smith), *Michael Bell (Doctor Octopus), Peter Cullen (Hulk), Vic Perrin (Thor), Michael Rye (Magneto),* John Haymer (Cyberiad), *Dennis Marks (Green Goblin), Christopher Collins (Sandman),* Marilyn Schreffler (Bonnie)

**Dir:** Don Jurwich, Bob Richardson; **Prod:** Dennis Marks; **Music:** Johnny Douglas; **Exec.Prod:** David H DePatie, Lee Gunther; **Marvel Productions/ Mihahn**; 25

**Notes:** Season two and the first four episodes of season three were broadcast with *The Incredible Hulk*, though they were edited together as a single episode called *The Incredible Hulk And The Amazing Spider-Man* for the second season, and then *The Amazing Spider-Man And The Incredible Hulk* for the second season. Stan Lee narrated the episodes as of the second season, but he then added narrations to the first season – though the current master tapes of the first two seasons don't have the narrations at all. Firestar was created especially for the program because of rights issues surrounding the Human Torch, who was the original choice; she would ultimately be incorporated into the comics, though via *The X-Men* rather than *Spider-Man*. A lot of continuity with the comics was kept surprisingly close, though the introduction of Lightwave (Aurora Dante) as Bobby Drake's half-sister was a much larger change. An urban legend suggested that this series was a sequel to *Spider-Man*, but as the two series were in production at the same time, and launched the same day, it's highly unlikely. That said, one episode features a flashback to an episode of *Spider-Man*, and the title cards are very similar. Strangely, while the production teams on both series were almost identical, the regular casts were very different. Many of the guest cast, however, did play the same characters.

**Ratings:** *IMDB:* 73%

**Review:** A step down from *Spider-Man*. There are a lot of nice touches, but dumbing down Ice Man can be a bit frustrating and the series does take on an *Archie* vibe from time to time. It's probably worth checking out, but if you have a choice between this and Spidey on his own, go with the latter.

## *THE INCREDIBLE HULK*
### *(September, 1982 – October, 1983/NBC)*

Bruce Banner attempts to find ways to stop his transformations, but the constant presence of villains such as Doctor Octopus, Hydra, the Leader and the Puppet Master mean the Hulk is required more often than not.

**Regular Cast:** *Michael Bell (Dr Bruce Banner), Pat Fraley (Maj Ned Talbot), Bob Holt (The Hulk), Michael Horton (Rick Jones), B J Ward (Betty Ross),* Stan Lee (Narrator)

**Episodes:**

*1.1 Tomb of the Unknown Hulk*

*1.2 Prisoner of the Monster*

*1.3 Origin of the Hulk*

*1.4 When Monsters Meet*

*1.5 The Cyclops Project*

*1.6 Bruce Banner Unmasked*

*1.7 The Creature and the Cavegirl*

*1.8 It Lives! It Grows! It Destroys!*

| | |
|---|---|
| *1.9 The Incredible Shrinking Hulk* | *2.3 The Boy Who Saw Tomorrow* |
| *2.1 Punks on Wheels* | *2.4 The Hulk Destroys Bruce* |
| *2.2 Enter: She-Hulk* | *Banner* |

**Notable Guest Cast:** Susan Blu (Rita), Hamilton Camp (Dr Brandon Jones), *Victoria Carroll (She-Hulk),* Ron Feinberg (Cosmos), June Foray (Umeela), *Stanley Jones (The Leader), Bob Ridgley (General Thunderbolt Ross), Stanley Ralph Ross (Quasimodo), Michael Rye (Supreme Hydra)*

**Prod:** Don Jurwich; **Music:** Johnny Douglas; **Exec.Prod:** David H DePatie, Lee Gunther; **Ed:** Robert T Gillis, Jeffrey L Sandler; **Marvel Productions/Pan Sang East Co**; 25

**Notes:** A spin-off of sorts from ***Spider-Man and His Amazing Friends***. The artwork was based on that done by Sal Buscema. Censorship was something of a problem for the series, due to the level of violence. One thing that wasn't, however, was that when turning back from the Hulk to Banner, Banner's clothes would reappear.

**Ratings:** *IMDB:* 69%

**Review:** Very much a product of its time, there is a certain charm about the series as a curio, but outside of that, it's very straightforward. Stan Lee's narrations are a little intrusive, but there's a few clever moments of dialogue all the same, and the decision to include Rita is actually a pretty good one. It's unlikely to be raved about, but there are worse shows in this list.

## *X-MEN: PRYDE OF THE X-MEN*
### *(16/9/1989)*

As Kitty Pryde starts her schooling at Xavier's School for Gifted Youngsters, she finds herself at the centre of an attack by Magneto on the school, in which she fails to protect a circuit of the Cerebro computer. When the X-Men go to attack Magneto, Kitty joins them to redeem herself.

**Regular Cast:** *Michael Bell (Cyclops), Earl Boen (Colossus), Andi Chapman (Storm), Pat Fraley (Pyro), Ronald Grans (Magneto), Allan Oppenheimer (Blob), Patrick Pinney (Juggernaut), Neil Ross (Wolverine), Susan Silo (White Queen), Kath Soucie (Kitty Pryde), John Stephenson (Professor X), Alexandra Stoddart (Dazzler), Frank Welker (Toad/Lockheed/Nightcrawler),* Stan Lee

(Narrator)

**Dir:** Ray Lee, Stu Rosen; **Writer:** Larry Parr; **Prod:** Rick Hoberg, Larry Houston, Will Meugniot; **Music:** Rob Walsh; **Exec.Prod:** Lee Gunther, Margaret Loesch Stimpson; **Ed:** Al Breitenbach; **Marvel Productions/New World Television**; 25

**Notes:** Marvel Productions' chief David H DePatie stepped down from his role after this production, marking this as the final animated Marvel series which involved either DePatie or Fitz Freleng (owners of DFE). The pilot was made with money left over from *RoboCop: The Animated Series* and was created by Toei Animation. Shortly thereafter, New World Pictures sold the Marvel Entertainment Group to the Andrews Group, and killed any chance of this going to series, though Loesch remained very interested in the product.

**Ratings:** *IMDB:* 69%

**Review:** The animation and design on this series are wonderful, but the voice acting is hideous, with some astonishingly bad choices (Wolverine is played as an Australian, though the accent flickers between that and Cockney). Additionally the plot is fairly straightforward, Stan Lee's narrations feel very old fashioned, and any character development is very basic.

## *SOLARMAN*
### *(24/10/1992/Fox Kids)*

Benjamin Tucker dreams of being an artist for Marvel Comics, but when an alien cyborg named Kraal attempts to drain the Sun's energy, Ben meets Kraal's rebel son Sha-han who gives him the Circlet of Power and a robot called Beepie. When Ben dons the Circlet, he becomes Solarman!

**Cast:** M G Kelly, Bernie Erhard, Stan Jones, Lou Hunt, Pat Fraley, Adam Carl, Angela Lee Sloan, Bettina Bush

**Dir:** John Gibbs; **Writers:** Chuck Menville, Len Janson; **Prod:** Jim Duffy; **Music:** Rob Walsh; **Exec.Prod:** Lee Gunther, Margaret Loesch, David Oliphant; **Marvel Productions**; 22

**Notes:** The most curious of curios. This pilot episode was actually made in 1986, but failed to impress enough for a series. Stan Lee and David Oliphant (the creator) got the comic series launched in 1989, at which point the episode was released on VHS to tie in, but after two issues the comic was cancelled. It was ultimately broadcast in 1992.

**Ratings:** *IMDB:* 56%

**Review:** It's not hard to see why this didn't get past the pilot. The animation is perfectly acceptable for the eighties (even the nineties, if you can forgive the eighties designs), but the story makes little sense, and it seems not to know whether it wants to be a parody of a superhero series, or an actual superhero series. As stated, it's a curio, but not really worth watching.

## *X-MEN*
### *(October, 1992 – September, 1997/Fox Kids)*

The X-Men – Professor X, Cyclops, Jean Grey, Wolverine, Rogue, Storm, Beast, Jubilee, Gambit and Morph – take on the threats to human/mutant relations, including the anti-mutant human bigots and their Sentinel Robots, mutant terrorist Magneto, and the powerful mutant Apocalypse who plans to enforce survival of the fittest.

**Regular Cast:** *Norm Spencer (Scott Summers/Cyclops), Cathal J Dodd (Wolverine), Lenore Zann (Rogue), Iona Morris & Alison Sealy-Smith (Ororo Munroe/Storm), George Buza (Hank McCoy/Beast), Chris Potter & Tony Daniels (Remy LaBeau/Gambit), Alyson Court (Jubilation Lee/Jubilee), Catherine Disher (Jean Grey/Phoenix), Cedric Smith (Professor Charles Xavier)*

**Episodes:**

*1.1 Night of the Sentinels, Part 1*

*1.2 Night of the Sentinels, Part 2*

*1.3 Enter Magneto*

*1.4 Deadly Reunions*

*1.5 Captive Hearts*

*1.6 Cold Vengeance*

*1.7 Slave Island*

*1.8 The Unstoppable Juggernaut*

*1.9 The Cure*

*1.10 Come the Apocalypse*

*1.11 Days of Future Past, Part 1*

*1.12 Days of Future Past, Part 2*

*1.13 The Final Decision*

*2.1 Till Death Us Do Part, Part 1*

*2.2 Till Death Us Do Part, Part 2*

*2.3 Whatever It Takes*

*2.4 Red Dawn*

*2.5 Repo Man*

*2.6 X-Ternally Yours*

*2.7 Time Fugitives, Part 1*

*2.8 Time Fugitives, Part 2*

*2.9 A Rogue's Tale*

**Notable Guest Cast:** Stephen Ouimette (Angel), John Colicos & James Blendick (Apocalypse), Jeremy Ratchford (Banshee), Philip Akin (Bishop), Rick Bennett (Juggernaut), Robert Cait (Colossus), Tracey Moore (Emma Frost), Marc Strange (Forge), Dennis Akayama (Iceman), Tasha Simms

(Psylocke), Jane Luk (Lady Deathstrike), Jennifer Dale (Domino), Randall Carpenter (Mystique), Tara Strong (Magik), Lally Cardeau (Moira MacTaggert), Ron Rubin (Morph), Paul Haddad (Nightcrawler), Susan Roman (Scarlet Witch), Terri Hawkes (Polaris)

**Prod:** Larry Houston; **Music:** Shuki Levy, Kussa Manchi, Ron Wasserman, Haim Saban; **Exec.Prod:** Avi Arad, Eric S Rollman, Joseph Calamari, Winston Richard, Stan Lee; **Prod.Des.:** John Petrovitz, Shannon Denton; **Ed:** Sharon Janis, John C Walts; **Genesis Entertainment/Graz Entertainment/Marvel Enterprises/Marvel Productions/Saban Entertainment**; 22

**Notes:** Margaret Loesch pushed this series through after her work on *Pryde of the X-Men*. The animation studio responsible for the show, AKOM, submitted the first two episodes with hundreds of animation errors which they refused to fix, forcing Fox to rethink their broadcast strategy. Ultimately they broadcast the first three episodes as a special event in 1992, before threatening AKOM with legal action to fix the episodes for a repeat broadcast the following year. Saban had to pay for the last six episodes alone because Marvel was filing for bankruptcy at the time; the animation was done by the Philippine Animation Studio for these episodes. This is the longest running Marvel animated television series, at a mammoth 76 episodes. The storylines were, for the most part, original, though some took inspiration and titles from the comics. The last six episodes were the Philippine episodes, but as of 3.8, the broadcast order for the series (as shown above) was very different to the intended order. The four part "Beyond Good and Evil" was originally intended to be the end of the series. The series was an enormous success, and Marvel created a comic based on this version.

**Ratings:** *IMDB:* 85%

**Review:** Given the ashes that this phoenix sprang from, it's a much better interpretation of the X-Men. The costumes are now in line with the nineties comics version, so personal preference will determine whether the designs are better or not, but the vocal work is much better, and there is a lot more character development with both X-Men and their opponents. This is largely remembered as *the* X-Men cartoon, and not without just cause.

## FANTASTIC FOUR

### (September, 1994 – February, 1996)

Reed Richards, along with his friend Ben, his lover Sue and her brother Johnny, gain superpowers from cosmic energy and use them to fight super powered villainy across the world.

**Regular Cast:** *Beau Weaver (Reed Richards/Mr Fantastic), Lori Alan (Susan Storm/Invisible Woman), Brian Austin Green & Quinton Flynn (Johnny Storm/ Human Torch), Chuck McCann (Ben Grimm/The Thing), John Vernon, Neil Ross & Simon Templeman (Dr Victor Von Doom), Pauline Arthur Lomas (Alicia Masters)*

**Episodes:**

*1.1 The Origin of the Fantastic Four, Part 1*

*1.2 The Origin of the Fantastic Four, Part 2*

*1.3 Now Comes the Sub-Mariner*

*1.4 Incursion of the Skrull*

*1.5 The Silver Surfer and the Coming of Galactus, Part 1*

*1.6 The Silver Surfer and the Coming of Galactus, Part 2*

*1.7 Super Skrull*

*1.8 The Mask of Doom, Part 1*

*1.9 The Mask of Doom, Part 2*

*1.10 The Mask of Doom, Part 3*

*1.11 Mole Man*

*1.12 Behold the Negative Zone*

*1.13 The Silver Surfer and the Return of Galactus*

*2.1 And a Blind Man Shall Lead Them*

*2.2 Inhumans Saga, Part 1: And the Wind Cries Medusa*

*2.3 Inhumans Saga, Part 2: The Inhumans Among Us*

*2.4 Inhumans Saga, Part 3: Beware the Hidden Land*

*2.5 Worlds Within Worlds*

*2.6 To Battle the Living Planet*

*2.7 Prey of the Black Panther*

*2.8 When Calls Galactus*

*2.9 Nightmare in Green*

*2.10 Behold, a Distant Star*

*2.11 Hopelessly Impossible*

*2.12 The Sentry Sinister*

*2.13 Doomsday*

**Notable Guest Cast:** *Robin Sachs & Edward Albert (Silver Surfer), Gregg Berger (Mole Man), Rocky Carroll (Triton), Michael Dorn (Gorgon), Richard Grieco (Ghost Rider), Mark Hamill (Maximus The Mad), Jess Harnell (Impossible Man), Tony Jay (Galactus),* Joan Lee (Mrs Lavinia Forbes), Stan Lee (Himself), *Alan Oppenheimer (Uatu), Ron Perlman (Bruce Banner/Hulk), John Rhys-Davies (Thor), Bill Smitrovich (Daredevil)*

**Prod:** Glen Hill, Dennis Ho, Paul B Strickland; **Music:** William Anderson, Giorgio Moroder; **Exec.Prod:** Avi Arad, Rick Ungar, Glen Kennedy, Stan Lee; **Ed:** Kerry Dean Williams; **Genesis Entertainment/Marvel Enterprises/ Marvel Productions/New World Entertainment Films**; 22

**Notes:** The first of the New World Entertainment Marvel series, and alongside *Iron Man*, this was a syndicated series. Kennedy Cartoons and Wang Film Productions created the animation for the first season, but as the program summary might suggest, the series was so shallow it was received terribly (though the plots did come from, and were faithful to, the comic series), and Marvel opted to go to the Philippine Animation Studio Inc for the second season. A number of changes were made to the design and the plots were changed to be more mature, but the series was still regarded as a failure and cancelled.

**Ratings:** *IMDB:* 64%

**Review:** The animation's not bad (though the Thing still looks ordinary) and some of the vocal work is really bad (say hello again, Thing). However a lot more thought has gone into the stories and the character design is much better than previous versions. A more worthwhile cartoon, though still not quite reaching the potential of Marvel's first family.

## *IRON MAN*
### *(September, 1994 – February, 1996)*
Billionaire industrialist Tony Stark dons powerful armour to become Iron Man, and do battle with threats against the entire world.

**Regular Cast:** *Robert Hays (Tony Stark/Iron Man), James Avery & Dorian Harewood (James Rhodes/War Machine), Ed Gilbert & Robert Ito (The Mandarin), Jim Cummings (MODOK), John Reilly (Clint Barton/Hawkeye), Katherine Moffat & Jennifer Darling (Scarlet Witch), Casey Defranco & Jennifer Hale (Julia Carpenter/Spider-Woman), James Warwick & Tom Kane (Century)*

**Episodes:**

***1.1 And the Sea Shall Give Up Its Dead***
***1.2 Rejoice! I Am Ultimo, Thy Deliverer***

**Notable Guest Cast:** *Philip Abbot (Nick Fury), Neil Dickson (Dreadknight), Linda Holdahl (Hypnotia), Chuck McCann (Blizzard), Neil Ross (Fin Fang Foom), Tony Steedman & Efrem Zimbalist Jr (Justin Hammer)*

**Prod:** Glen Hill, Dennis Ho, Ted Tchoe [1.1-1.13]; **Music:** William Anderson, Keith Emerson; **Exec.Prod:** Avi Arad, Rick Ungar, Stan Lee; **Ed:** Catherine MacKenzie; **Marvel Enterprises/Marvel Productions/New World Entertainment Films**; 26

**Notes:** The second of the Marvel animated series to be syndicated, alongside **The Fantastic Four**. The program was originally created by the Rainbow Animation Group, but the reaction to the series was so poor thanks to the shallow storylines (again, we weren't skimping on the program summary), that Marvel opted to change to Koko Enterprises for the second season, which resulted in massive changes, including a change of mood and design, alongside a lot of voice recasting. Nonetheless, the change didn't help the series and the ratings declined, ultimately earning it cancellation.

**Ratings:** *IMDB:* 68%

**Review:** Despite the somewhat dodgy designs (Tony's mullet is very off putting), the stories are a lot better in the second season, and the series does improve a fair bit. The fact that the series uses the frequently underused Carpenter Spider-Woman is quite nice. It's worth skipping over the first season and watching the second.

## SPIDER-MAN

### (November, 1994 – January, 1998/Fox)

Having become the spectacular Spider-Man, Peter Parker balances his life in college with fighting crime and realising he's part of a much larger group of super powered heroes who are tackling equally powered villains.

**Regular Cast:** *Christopher Daniel Barnes (Peter Parker/Spider-Man), Sara Ballantine (Mary Jane Watson), Gary Imhoff (Harry Osborn/Green Goblin), Joseph Capanella (Dr Curt Connors/The Lizard), Jennifer Hale (Felicia Hardy/ Black Cat), Julie Bennett (Aunt May),* Susan Beaubian (Dr Mariah Crawford), *Liz Georges (Debra Whitman), Patrick Labyorteaux (Flash Thompson)*

**Episodes:**

*1.1 Night of the Lizard*
*1.2 The Spider Slayer*
*1.3 Return of the Spider Slayer*
*1.4 Doctor Octopus: Armed and Dangerous*
*1.5 The Menace of Mysterio*
*1.6 The Sting of the Scorpion*
*1.7 Kraven the Hunter*
*1.8 The Alien Costume, Part 1*
*1.9 The Alien Costume, Part 2*
*1.10 The Alien Costume, Part 3*
*1.11 The Hobgoblin, Part 1*
*1.12 The Hobgoblin, Part 2*
*1.13 Day of the Chameleon*
*2.1 The Insidious Six*
*2.2 Battle of the Insidious Six*
*2.3 Hydro-Man*
*2.4 The Mutant Agenda*
*2.5 Mutants' Revenge*
*2.6 Morbius*
*2.7 Enter the Punisher*
*2.8 Duel of the Hunters*
*2.9 Blade, the Vampire Hunter*
*2.10 The Immortal Vampire*
*2.11 Tablet of Time*

*2.12 Ravages of Time*
*2.13 Shriek of the Vulture*
*2.14 The Final Nightmare*
*3.1 Doctor Strange*
*3.2 Make a Wish*
*3.3 Attack of the Octobot*
*3.4 Enter the Green Goblin*
*3.5 Rocket Racer*
*3.6 Framed*
*3.7 The Man Without Fear*
*3.8 The Ultimate Slayer*
*3.9 Tombstone*
*3.10 Venom Returns*
*3.11 Carnage*
*3.12 The Spot*
*3.13 Goblin War!*
*3.14 Turning Point*
*4.1 Guilty*
*4.2 The Cat*
*4.3 The Black Cat*
*4.4 The Return of Kraven*
*4.5 Partners*
*4.6 The Awakening*
*4.7 The Vampire Queen*
*4.8 The Return of the Green Goblin*

**Notable Guest Cast:** *Edward Asner (J Jonah Jameson), Rodney Saulsberry (Robbie Robertson), Nell Carter (Glory Grant), Lauren Tom (Dr Chong-Yu), Eddie Albert (Vulture), Hank Azaria (Venom), Leigh-Allyn Baker (Alisha Silvermane), Majel Barrett (Anna Watson), Gregg Berger (Kraven), Nicky Blair (Hammerhead), Roscoe Lee Browne (Kingpin), Scott Cleverdon (Carnage), Jeff Corey (Silvermane), Jim Cummings (Chameleon), Ed Gilbert (Dormammu), Mark Hamill (Hobgoblin), David Warner (Dr Landon), Dorian Harewood (Tombstone), Nick Jameson (Morbius), Tony Jay (Baron Mordo), Martin Landau (Scorpion), Joan Lee (Madame Webb), Mira Furlan (Silver Sable), David Hayter (Captain America), Tom Kane (Doctor Doom),* Stan Lee (Himself), Rue McClanahan (Anastacia Hardy), Malcolm McDowell (Whistler), *Tim Russ (Prowler), George Takei (Wong)*

**Dir:** Robert Shellhorn [1.1], Bob Richardson [1.2-5.13]; **Prod:** Stanley Liu, John Semper, Koji Takeuchi, Joseph Lampone Jr; **Music:** Shuki Levi, Haim Saban; **Exec.Prod:** Avi Arad, Stan Lee; **Prod.Des.:** Vladimir Spasojevic; **Ed:** George D Brown III, Mark Deimel; **Genesis Entertainment/Marvel Enterprises/ Marvel Productions/New World Entertainment Films/TMS Entertainment**; 21

**Notes:** An enormously successful television series, the program met an unexpected end when the 65 episode contract was met. The program was created by Marvel Films Animation, a company headed by Avi Arad. Network boss Margaret Loesch hated Arad so much she wanted to drive him out of business and subsequently cancelled the program. Toy lines, video games and comic series were all spawned from the program. Season Two had the title "Neogenic Nightmare"; season three had "Sins of the Father", while season four was "Partners in Danger". The program ends on a cliff-hanger because the producers were convinced that they would get another season, and were unaware of the Loesch/Arad dispute.

**Ratings:** *IMDB:* 83%

**Review:** As good as its reputation suggests, the use of much of the comic series mythology as well as a fairly impressive cast makes the series very watchable. The design work is still very nineties, but the writing and acting make up for some of the more dated aspects, and it's definitely a series worth watching.

## *THE INCREDIBLE HULK*
### *(September, 1996 – November, 1997/UPN)*

On the run from General Ross and Major Talbot, Bruce Banner also finds himself the target of the Leader and his gamma warriors, with nothing but help from Rick Jones and Betty Ross. Betty and Doc Samson work to find a way to stop Bruce's transformations, but when Doctor Doom wounds Bruce's cousin Jennifer Walters, a blood transfusion sees her become She-Hulk, while Samson's work changes Rick's life, and alters Bruce's transformation, rather than stopping it.

**Regular Cast:** *Neal McDonough (Dr Robert Bruce Banner), Lou Ferrigno (The Hulk), Genie Francis & Philece Sampler (Betty Ross), Michael Donovan (Grey Hulk), Luke Perry (Rick Jones), John Vernon (General Thunderbolt Ross), Kevin Schon (Major Talbot), Matt Frewer (Leader), Mark Hamill (Gargoyle), Shadoe Stevens (Doc Samson), Lisa Zane & Cree Summer (She-Hulk)*

**Episodes:**

**1.1 Return of the Beast, Part 1**
**1.2 Return of the Beast, Part 2**
**1.3 Raw Power**
**1.4 Helping Hand, Iron Fist**
**1.5 Innocent Blood**
**1.6 Man to Man, Beast to Beast**
**1.7 Doomed**
**1.8 Fantastic Fortitude**
**1.9 Mortal Bounds**

**1.10 And The Wind Cries... Wendigo!**
**1.11 Darkness and Light, Part 1**
**1.12 Darkness and Light, Part 2**
**1.13 Darkness and Light, Part 3**
**2.1 Hulk of a Different Color**
**2.2 Down Memory Lane**
**2.3 Mind Over Anti-Matter**
**2.4 They Call Me Mr Fixit**

| | |
|---|---|
| *2.5 Fashion Warriors* | *2.7 The Lost Village* |
| *2.6 Hollywood Rocks* | *2.8 Mission: Incredible* |

**Notable Guest Cast:** *Kathy Ireland (Ogress), Richard Moll (Abomination), Thom Barry (Izzy Cohen), Simon Templeman (Dr Doom), Beau Weaver (Mr Fantastic), Robert Hays (Iron Man), Dorian Harewood (War Machine), Richard Grieco (Ghost Rider), Jim Cummings (Shocker), Maurice LaMarche (Dr Strange), John Rhys-Davies (Thor), Mark L Taylor (Donald Blake), Chuck McCann (The Thing),* Kevin Michael Richardson (Dark Hulk), Stan Lee (Mr Walters)

**Prod:** Dick Sebast, Tom Tataranowicz, Ron Myrick; **Music:** Shuki Levi, Haim Saban; **Exec.Prod:** Avi Arad, Rick Ungar, Eric S Rollman, Stan Lee; **Ed:** Catherine MacKenzie [1.1-1.13], Jonathon Braun [2.1-2.8]; **Genesis Entertainment/Marvel Enterprises/Marvel Productions/New World Entertainment Films**; 22

**Notes:** The fourth of the New World Entertainment/Marvel series; the Hulk had appeared in both *Iron Man* and *Fantastic Four* previously in the hope he would work out. Unfortunately Marvel were not impressed with the first season, and requested a change of production personnel, which failed to make much of a difference. As of the second season, the program was retitled *The Incredible Hulk and She-Hulk*.

**Ratings:** *IMDB:* 68%

**Review:** It's probably worth noting that there is a very different feel between seasons, with the second season far lighter and more humorous, and the Hulk getting pushed back a bit. The first season is darker and moodier, but it does work on a few levels. The second season removes that darkness, but arguably isn't the better for it.

## SILVER SURFER
### *(February – May, 1998/Fox Kids)*

Norin Radd gives himself up to be the Herald for the world eater Galactus in order to save the life of his planet Zenn-La. When Thanos uses the living planet Ego to find Galactus' secrets in the Surfer's mind, he instead restores the Surfer's free will, and as Galactus is about to gorge on Earth, the Surfer rises up against him.

**Regular Cast:** *Paul Essiembre (Norin Radd), James Blendick (Galactus)*

**Episodes:**

*1.1 The Origin of the Silver Surfer, Part 1*

*1.2 The Origin of the Silver Surfer, Part 2*

*1.3 The Origin of the Silver Surfer, Part 3*

*1.4 The Planet of Dr Moreau*

*1.5 Learning Curve, Part 1*

*1.6 Learning Curve, Part 2*

*1.7 Innervisions*

*1.8 Antibody*

*1.9 Second Foundation*

*1.10 Radical Justice*

*1.11 The Forever War*

*1.12 Return to Zenn-La*

*1.13 The End of Eternity, Part 1*

**Notable Guest Cast:** *Camilla Scott (Shalla-Bal), John Neville (Eternity), Tara Rosling (Nova), Elizabeth Shepherd (Infinity), Gary Krawford (Thanos), Colin Fox (Uatu), Alison Sealy-Smith (Gamora), Norm Spencer (Drax), Jennifer Dale (Nebula), Roy Lewis (Ego), Oliver Becker (Adam Warlock), Karl Pruner (Beta Ray Bill)*

**Dir:** Roy Allen Smith; **Prod:** Dale Hendrickson, Roy Allen Smith, Tom McLaughlin; **Music:** Ian Christian Nickus; **Exec.Prod:** Avi Arad, Eric S Rollman, Stan Lee; **Ed:** John C Walts; **Saban Entertainment/Marvel Entertainment**; 25

**Notes:** A number of scripts were written to continue the series, including a conclusion to the final episode, and all written by Larry Brody. A deliberate attempt was made with the animation to recreate Jack Kirby's comic style for the series; an attempt which was relatively successful. One of three series that Saban were making with Marvel, it's not entirely clear why all three were cancelled, though Brody claims that a legal dispute was the cause for the end of this one. Marvel's financial situation at the time no doubt also played a part.

**Ratings:** *IMDB:* 71%

**Review:** The Silver Surfer is a po-faced hero at the best of times, and this is completely true here. A Nova series would have been much more exciting, as she's far more interesting in this series. The Kirby-style art is fantastic, but oddly the Galactus scenes seem strangely out of place. It's not a terrible series, but it's quite boring, in truth.

## SPIDER-MAN UNLIMITED
### (October, 1999 – March, 2001/Fox Kids)

When John Jameson heads to Counter-Earth, Spider-Man is forced to stop Venom and Carnage from disrupting the mission, but when Jameson loses contact with Earth, it is Spider-Man who is blamed. Believed to be dead by the world after saving innocents in a fire, Spider-Man gets a new suit from Reed Richards and heads to Counter-Earth to rescue Jameson.

**Regular Cast:** *Rino Romano (Peter Parker/Spider-Man),* Akiko Morrison (Dr Naoko Yamada-Jones), *Paul Dobson (The Hunter),* Kathleen Barr (Prima), *Dale Wilson (Machine Man),* Richard Newman (High Evolutionary), *John Payne (John Jameson)*

**Episodes:**

*1.1 Worlds Apart, Part One*
*1.2 Worlds Apart, Part Two*
*1.3 Where Evil Nests*
*1.4 Deadly Choices*
*1.5 Steel Cold Heart*
*1.6 Enter The Hunter!*
*1.7 Cry Vulture*

*1.8 Ill-Met By Moonlight*
*1.9 Sustenance*
*1.10 Matters of the Heart*
*1.11 One Is the Loneliest Number*
*1.12 Sins of the Father*
*1.13 Destiny Unleashed*

**Notable Guest Cast:** Kim Hawthorne (Karen O'Malley), Christopher Gaze (Daniel Bromley), *Brian Drummond (Venom),* David Sobolov (Lord Tyger), Ron Halder (Sir Ram), Tasha Simms (Lady Ursula), *Jennifer Hale (Mary Jane Watson),* Scott McNeil (Vulture)

**Dir:** Patrick Archibald; **Prod:** Will Meugniot; **Music:** Ian Christian Nickus; **Exec.Prod:** Avi Arad, Eric S Rollman; **Saban Entertainment/Marvel Entertainment**; 22

**Notes:** Another curious cartoon, which had the most bizarre broadcast (the first three episodes were broadcast in 1999, the fourth in December, 2000, and the other nine in 2001). There were talks about crossing this series over with *The Avengers: United They Stand*, but ultimately that never happened. Due to the deal with Sony, Saban were unable to use any material from the comics, and so the intention to adapt *The Amazing Spider-Man* was dropped, and Spider-Man was given a new costume. Pokemon effectively killed this series as Saban felt it wasn't worthwhile pursuing it.

**Ratings:** *IMDB:* 66%

**Review:** This is a really odd cartoon. Viewers are expected to know pretty much the entire Spider-Man backstory, as they are thrown into an adventure on a different world. And yet...this has the best title sequence of any Spider-Man cartoon, has some fantastic design work (the updated Spider-suit is very reminiscent of the Scarlet Spider), and really good voice work. It's a great cartoon, but not one for newcomers.

### THE AVENGERS: UNITED THEY STAND
#### *(October, 1999 – February, 2000/Fox Kids)*

The Avengers fight villains that threaten our way of life, using their suits of battle armour. (Yes, you read that last bit correctly.)

**Regular Cast:** *Linda Ballentyne (Wasp), Tony Daniels (Hawkeye), Ron Rubin (Vision), Martin Roach (Falcon), Rod Wilson (Ant-Man), Lenore Zann (Tigra), Stavroula Logothettis (Scarlet Witch), Hamish McEwan (Wonder Man), Graham Harley (Jarvis), Ray Landry (Raymond Sikorsky), Gerry Mendicino (Taurus), John Stocker (Ultron),* Michael Yarmush (Announcer)

**Episodes:**

| | |
|---|---|
| *1.1 Avengers Assemble, Part One* | *1.8 Shooting Stars* |
| *1.2 Avengers Assemble, Part Two* | *1.9 What a Vision Has to Do* |
| *1.3 Kang* | *1.10 Egg-Streme Vengeance* |
| *1.4 Comes a Swordsman* | *1.11 The Sorceress's Apprentice* |
| *1.5 Remnants* | *1.12 Earth and Fire, Part One* |
| *1.6 Command Decision* | *1.13 Earth and Fire, Part Two* |
| *1.7 To Rule Atlantis* | |

**Notable Guest Cast:** *Philip Akin (Attuma), Oliver Becker (Absorbing Man), Dan Chameroy (Captain America), Francis Diakowsky (Iron Man), Ken Kramer (Kang), Robert Latimer (Egghead), Susan Roman (Moonstone), Allan Royal (Grim Reaper), Elizabeth Shepherd (Agatha Harkness), Phillip Shepherd (Baron Zemo), Raoul Trujillo (Namor)*

**Dir/Prod:** Ron Myrick; **Music:** Shuki Levy, Haim Saban; **Exec.Prod:** Avi Arad, Eric S Rollman, Stan Lee; **Film Roman Productions/Marvel Entertainment**; 22

**Notes:** A somewhat bizarre series that saw the Avengers (essentially the line-up from the comic series *The West Coast Avengers*) operating twenty-five years in the future, inspired strongly by the popular **Batman Beyond** cartoon. The four most notable Avengers were missing, Eric Lewald (script editor) claiming that they wanted the series to focus on the Avengers, not a specific character, though in truth it was more likely due to licensing issues. Captain America and Iron Man do appear in cameo, and Thor and the Hulk were planned for season two, which would also have seen an appearance by the X-Men. Ratings and reviews were not kind, however, and the series was cancelled. A comic series based on this was also produced, but quickly went the way of the series.

**Ratings:** *IMDB:* 54%

**Review:** Whilst the animation for the series isn't particularly bad, in truth the series is quite cheesy and has some very awful voice acting (Scarlet Witch is unbearable!). Although it's nice to see a different line up of Avengers – and under the command of Ant-Man, no less – the whole "armoured suits" gimmick cheapens the team, and there's little to recommend it.

## X-MEN: EVOLUTION
### (November, 2000 – October, 2003/Kids WB)

Charles Xavier and Magneto have different visions for how mutants should exist in society, and both are keen to recruit mutants to their way of thinking. As Xavier's X-Men train to take on Magneto's Brotherhood of Mutants, the American government starts the Sentinel program and Apocalypse sits to battle both sides.

**Regular Cast:** *Kirby Morrow (Scott Summers), Venus Terzo (Jean Grey), David Kaye (Professor Xavier), Brad Swaile (Nightcrawler), Maggie Blue O'Hara (Kitty Pryde), Scott McNeil (Logan), Meghan Black (Rogue), Kirsten Williamson (Ororo Munroe),* Neil Dennis (Evan Daniels), *Michael Kopsa (Hank McCoy)*

**Episodes:**

*1.1 Strategy X*
*1.2 The X-Impulse*
*1.3 Rogue Recruit*
*1.4 Mutant Crush*
*1.5 Speed and Spyke*
*1.6 Middleverse*
*1.7 Turn of the Rogue*
*1.8 SpykeCam*
*1.9 Survival Of The Fittest*
*1.10 Shadowed Past*

**Notable Guest Cast:** *Noel Fisher (Toad), Christopher Gray (Avalanche), Michael Dobson (Blob), Andrew Francis (Iceman), Colleen Wheeler (Mystique), Richard Ian Cox (Quicksilver), Christopher Judge (Magneto), Alexandra Carter (Amara), Tony Sampson (Berserker), Bill Switzer (Cannonball), Michael Donovan (Sabretooth), Megan Leitch (Boom Boom), Kelly Sheridan (Scarlet Witch)*

**Prod:** Boyd Kirkland, Greg Johnson, Michael Wolf; **Music:** William Anderson; **Exec.Prod:** John Bush, John W Hyde, John F Vein, Avi Arad, Stan Lee, Rick Ungar; **Film Roman Productions/Marvel Entertainment**; 22

**Notes:** The release of **The X-Men** influenced the design of the second season, which made Kirkland happy as he didn't really like the design of the characters for the first season. The producers and the network frequently were at loggerheads in regards to the direction of the show as the network wanted a far more childish program. As of the third season, the producers deliberately took the show down a darker route (and got a new title sequence). The fourth season was cut short, with a scene at the end of the ninth episode suggesting a variety of futures. The production team had hoped to develop the Dark Phoenix storyline before it was cancelled.

**Ratings:** *IMDB:* 79%

**Review:** An example of a television show that was compromised on its artistic vision by a network that had a very different one. The result is something that doesn't entirely work, though as the series continues to gain confidence it becomes better as a result. There are some nice ideas and the animation works well, but overall it doesn't quite come up to scratch.

## *SPIDER-MAN: THE NEW ANIMATED SERIES*
### *(July – September, 2003/MTV)*

With Norman Osborn dead, Peter Parker attends Empire State University alongside Mary Jane and Harry, and tries to build a relationship with Mary Jane that is complicated by his life as Spider-Man.

**Regular Cast:** *Neil Patrick Harris (Peter Parker/Spider-Man), Lisa Loeb (Mary Jane Watson), Ian Ziering (Harry Osborn)*

**Episodes:**

| | |
|---|---|
| *1.1 Heroes and Villains* | *1.8 When Sparks Fly* |
| *1.2 The Sword of Shikata* | *1.9 Royal Scam* |
| *1.3 Keeping Secrets* | *1.10 The Party* |
| *1.4 Tight Squeeze* | *1.11 Flash Memory* |
| *1.5 Law of the Jungle* | *1.12 Mind Games Part 1* |
| *1.6 Head over Heels* | *1.13 Mind Games Part 2* |
| *1.7 Spider-Man Dis-Sabled* | |

**Notable Guest Cast:** *Keith Carradine (J Jonah Jameson), Rob Zombie (Curt Connors/Lizard),* Eve (Talon), *Michael Dorn (Kraven), Michael Clarke Duncan (The Kingpin), Virginia Madsen (Silver Sable),* James Marsters (Sergei), John C McGinley (Richard Damian), *Ethan Embry (Max Dillon/Electro), Devon Sawa (Flash Thompson),* Kathy Griffin (Roxanne Gaines), Jeremy Piven (Roland Gaines), Stan Lee (Frank Elson)

**Prod:** Steven Wedland, Barbara Zelinski; **Music:** William Anderson; **Exec.Prod:** Avi Arad, Stan Lee, Morgan Gendel, Rick Ungar; **Adelaide**

Productions/Sony Pictures Television/Mainframe Entertainment/Marvel Entertainment; 22

**Notes:** For no apparent reasons, the broadcast order of episodes as listed, is clearly not the intended order. The series seems to follow the Sam Raimi **Spider-Man** movie chronology (though the casting of Michael Clarke Duncan hints it may also acknowledge those movies as well). A number of guest cast have appeared in regular roles from either earlier Spider-Man animated series, or would go onto appear regularly in future Spider-Man animated series. The series was broadcast on MTV, aimed at a younger audience, but the ratings weren't enough to guarantee a second season.

**Ratings:** *IMDB:* 72%

**Review:** A well written series, with an impressive guest cast, but the style of animation is very unique and is very much a "love-it-or-hate-it" sort of style. Honestly, I don't like it and the rendering of some of the characters – particularly Electro – make for a disjointed tone.

## *FANTASTIC FOUR: WORLD'S GREATEST HEROES*
### *(September, 2006 – October, 2007/Cartoon Network)*
The Fantastic Four work to defeat the villainy that threatens the world, working alongside other super heroes and confronting the villainous Dr Victor Von Doom.

**Regular Cast:** *Hiro Kanagawa (Reed Richards/Mr Fantastic), Lara Gilchrist (Susan Storm/Invisible Woman), Christopher Jacot (Johnny Storm/Human Torch), Brian Dobson (Ben Grimm/The Thing), Samuel Vincent (HERBIE), Paul Dobson (Victor Von Doom/Dr Doom), Sunita Prasad (Alicia Masters)*

**Episodes:**

| | |
|---|---|
| *1.1 Doomsday* | *1.8 My Neighbour Was a Skrull* |
| *1.2 Molehattan* | *1.9 World's Tiniest Heroes* |
| *1.3 Trial by Fire* | *1.10 De-Mole-Ition* |
| *1.4 Doomed* | *1.11 Impossible* |
| *1.5 Puppet Master* | *1.12 Bait & Switch* |
| *1.6 Zoned Out* | *1.13 Annihilation* |
| *1.7 Hard Knocks* | *1.14 Revenge of the Skrulls* |

**Notable Guest Cast:** *Michael Adamthwaite (Namor), Don Brown (Peter Henry Gyrich), Trevor Devall (Diablo), Michael Dobson (Ronan), Andrew Kavadas (Bruce Banner), Johnathan Holmes (The Wizard), David Kaye (Tony Stark/Iron Man), Terry Klassen (Impossible Man), Scott McNeil (Annihilus), Colin Murdock (Willie Lumpkin), John Payne (Henry Pym/Ant-Man), Alvin Sanders (Puppet Master), Rebecca Shoichet (Jennifer Walters/She-Hulk), Samuel Vincent (Trapster)*

**Dir:** Franck Michel; **Prod:** Benoît di Sabatino, Christophe di Sabatino; **Exec.Prod:** Craig Kyle, Eric S Rollman, Avi Arad, Nicolas Atlan; **Moonscoop/ Marvel Enterprises**; 22

**Notes:** An interesting version of the Fantastic Four which attempts to take elements from all the different versions that had been done to that point, including the *Ultimate Fantastic Four* comics, and some of the earlier cartoons (note HERBIE). The stories, however, were mostly original. Poor ratings, due in part to terrible scheduling, meant that a second season was not made.

**Ratings:** *IMDB:* 69%

**Review:** A really nice take on the FF, particularly the design of the costumes which nods in some ways to the Future Foundation costumes. The characters are brought to life well, and the series makes good use of the Marvel universe, and in particular the various elements from the *Fantastic Four* comics. It's surprising that a second season wasn't commissioned, though the Fantastic Four are probably the least popular of Marvel's properties regardless of the quality of the material.

## THE SPECTACULAR SPIDER-MAN
### (March, 2008 – November, 2010/Cartoon Network)

After being bitten by a radioactive spider and assuming the superhero identity of Spider-Man, Peter Parker juggles his crime fighting life with his real life, helping to support his Aunt May, working at the Daily Bugle and as a lab assistant to Dr Connors. However, Spider-Man has drawn the ire of "The Big Man" who runs crime in New York, and another criminal has arrived to take that position – The Green Goblin.

**Regular Cast:** *Josh Keaton (Peter Parker/Spider-Man), Lacey Chabert (Gwen Stacy), Benjamin Diskin (Eddie Brock/Venom), James Arnold Taylor (Harry Osborn), Alanna Ubach (Liz Allan), Kevin Michael Richardson (Tombstone), Daran Norris (J Jonah Jameson), Vanessa Marshall (Mary Jane Watson), Steven Blum (Green Goblin), Joshua LeBar (Flash Thompson), Andrew Kishino (Kenny Kong), Clancy Brown (Rhino), Alan Rachins (Norman Osborn)*

**Episodes:**

| | |
|---|---|
| *1.1 Survival Of The Fittest* | *2.1 Blue Prints* |
| *1.2 Interactions* | *2.2 Destructive Testing* |
| *1.3 Natural Selection* | *2.3 Reinforcement* |
| *1.4 Market Forces* | *2.4 Shear Strength* |
| *1.5 Competition* | *2.5 First Steps* |
| *1.6 The Invisible Hand* | *2.6 Growing Pains* |
| *1.7 Catalysts* | *2.7 Identity Crisis* |
| *1.8 Reaction* | *2.8 Accomplices* |
| *1.9 The Uncertainty Principle* | *2.9 Probably Cause* |
| *1.10 Persona* | *2.10 Gangland* |
| *1.11 Group Therapy* | *2.11 Subtext* |
| *1.12 Intervention* | *2.12 Opening Night* |
| *1.13 Nature Vs Nurture* | *2.13 Final Curtain* |

**Notable Guest Cast:** *Thom Adcox (Tinkerer), Edward Asner (Uncle Ben), Deborah Strang (Aunt May), Dee Bradley Baker (Curt Connors/Lizard), Irene Bedad (Jean DeWolff), Xander Berkeley (Mysterio), Nikki Cox (Silver Sable), John DiMaggio (Sandman), Robert Englund (Vulture), Miguel Ferrer (Silvermane), Crispin Freeman (Electro), Tricia Helfer (Black Cat), Phil LaMarr (Robbie Robertson), Stan Lee (Stan), Jane Lynch (Joan Jameson), Peter MacNicol (Dr Octopus), James Remar (Walter Hardy), Danny Trejo (Ox), Courtney B Vance (Roderick Kinglsey)*

**Super.Dir:** Victor Cook; **Prod:** Diane A Crea; **Music:** Kristopher Carter, Michael McCuistion, Lolita Ritmanis; **Exec.Prod:** Craig Kyle, Stan Lee, Eric S Rollman, Avi Arad; **Adelaide Productions/Culver Entertainment/Marvel Enterprises/Sony Pictures Television**; 22

**Notes:** Taking its nod from Sony's own Spider-Man movies, this series reworks that, as well as incorporating a number of other plot points and characters from various versions of the character. It was quite successful and a third season was planned, with Marina Sirtis even being approached for guest vocals. However, Sony surrendered their television rights and Disney acquired Marvel, both of which meant this series was no longer viable, and so it ended.

**Ratings:** *IMDB:* 81%

**Review:** A superb take on Spider-Man, with some excellent voice casting and clever story ideas. The art style is unique (sort of *Ben 10* really), but that works well for the target audience. The character design is very faithful to the comics as well. Definitely one of the better Spider-Man animated series.

## *WOLVERINE AND THE X-MEN*
### *(January – November, 2009/Nicktoons)*

One year after an explosion at Xavier's School for Gifted Youngsters appears to kill Xavier and Jean, Wolverine seeks out Beast to rebuild the X-Men. When Emma Frost approaches them, claiming she knows where Xavier actually is, and then does take them to his comatose body in Genosha, Wolverine is contacted by Xavier twenty years in the future. Xavier needs Wolverine to reunite all the X-Men in order to stop the creation of the Sentinel Master Mold, and the future that will come from that.

**Regular Cast:** *Steven Blum (Wolverine), Kieren van den Blink (Rogue), Susan Dalian (Storm), Jennifer Hale (Jean Grey), Danielle Judovits (Shadowcat), Tom Kane (Magneto), Yuri Lowenthal (Iceman), Nolan North (Cyclops), Liam O'Brien (Angel/Nightcrawler), Roger Craig Smith (Forge), Fred Tatasciore (Beast), Kari Wahlgren (Emma Frost), Jim Ward (Professor X)*

**Episodes:**

*1.1 Hindsight (Part 1)*       *1.3 Hindsight (Part 3)*
*1.2 Hindsight (Part 2)*       *1.4 Overflow*

**Notable Guest Cast:** *Charlie Adler (Mojo), Tamara Bernier (Mystique), Clancy Brown (Mr Sinister), Grey DeLisle (Psylocke), Alex Désert (Nick Fury), Kate Higgins (Scarlet Witch), Mark Hildreth (Quicksilver), Michael Ironside (Col Moss), Phil LaMarr (Gambit), Peter Lurie (Sabretooth), Gabriel Mann (Bruce Banner), Graham McTavish (Sebastian Shaw), Phil Morris (Colossus), Liza del Mundo (Polaris), Kevin Michael Richardson (Shadow King), Crystal Scales (Boom Boom), André Sogliuzzo (Arclight), Stephen Stanton (Blob), April Stewart (Selene), Tara Strong (Stepford Cuckoos), Gwendoline Yeo (Domino)*

**Prod:** Jason Netter; **Music:** Dean Grinsfelder; **Exec.Prod:** Stan Lee, Eric S Rollman, Avi Arad, Ed Borgerding, P Jayakumar; **Created by** Craig Kyle, Greg Johnson; **Marvel Enterprises/Toonz Animation/First Serve Toonz/Kickstart Productions/Liberation Entertainment**; 23

**Notes:** This series was based loosely on the *Astonishing X-Men* comic series. Despite excellent ratings and pre-production work for a second series, it was never renewed. Some have suggested that this was a deliberate move by Marvel against Fox, though this seems a little unlikely, as Fox wasn't benefitting from its existence. Many of the regular cast would voice their characters in other Marvel animated series.

**Ratings:** *IMDB:* 81%

**Review:** Easily the best animated X-Men series made, this one takes the series from a unique viewpoint (Wolverine being the focus, and the X-Men attempting to rebuild), and then brings in a wide variety of storylines from the comic series, including a spectacularly developed Phoenix Saga. With the setup for an Apocalypse storyline in season two, why this was never continued has me at a loss.

## IRON MAN: ARMORED ADVENTURES
### (April, 2009 – July, 2012/Nicktoons)

Sixteen year old Tony Stark is curious about the death of his father, and is convinced that Howard Stark's business partner Obadiah Stane is behind the death. Using armour he has built, Tony becomes Iron Man, and sets out to prove Stane's guilt, with the assistance of his friend James Rhodes, and girlfriend Pepper Potts.

**Regular Cast:** *Adrian Petriw (Tony Stark/Iron Man), Daniel Bacon (James Rhodes/War Machine), Anna Cummer (Pepper Potts/Rescue), Vincent Tong (Xin Zhang/Mandarin), Mackenzie Gray (Obadiah Stane (Iron Monger)*

**Episodes:**

*1.1 Iron, Forged In Fire: Part 1*
*1.2 Iron, Forged In Fire: Part 2*
*1.3 Secrets and Lies*
*1.4 Cold War*
*1.5 Whiplash*
*1.6 Iron Man Vs the Crimson Dynamo*
*1.7 Meltdown*
*1.8 Field Trip*
*1.9 Ancient History 101*
*1.10 Ready, AIM, Fire*
*1.11 Seeing Red*
*1.12 Masquerade*
*1.13 Hide and Seek*
*1.14 Man and Iron Man*
*1.15 Panther's Prey*
*1.16 Fun with Lasers*
*1.17 Chasing Ghosts*
*1.18 Pepper Interrupted*
*1.19 Technovore*
*1.20 World on Fire*
*1.21 Designed Only For Chaos*
*1.22 Don't Worry, Be Happy*
*1.23 Uncontrollable*
*1.24 Best Served Cold*

*1.25 Tales of Suspense: Part 1*
*1.26 Tales of Suspense: Part 2*
*2.1 The Invincible Iron Man, Part 1: Disassembled*
*2.2 The Invincible Iron Man, Part 2: Reborn!*
*2.3 Look into the Light*
*2.4 Ghost in the Machine*
*2.5 Armor Wars*
*2.6 Line of Fire*
*2.7 Titanium Vs Iron*
*2.8 The Might of Doom*
*2.9 The Hawk and the Spider*
*2.10 Enter: Iron Monger*
*2.11 Fugitive of SHIELD*
*2.12 All the Best People Are Mad*
*2.13 Heavy Mettle*
*2.14 Mandarin's Quest*
*2.15 Hostile Takeover*
*2.16 Extremis*
*2.17 The X-Factor*
*2.18 Iron Man 2099*
*2.19 Control-Alt-Delete*
*2.20 Doomsday*
*2.21 The Hammer Falls*

| | |
|---|---|
| **2.22 Rage of the Hulk** | **Annihilate!** |
| **2.23 Iron Monger Lives** | **2.26 The Makluan Invasion, Part 2:** |
| **2.24 The Dragonseed** | **Unite!** |
| **2.25 The Makluan Invasion, Part 1:** | |

**Notable Guest Cast:** Alistair Abell (Happy Hogan), Ashleigh Ball (Black Widow), Jeffrey Bowyer-Chapman (Black Panther), Christopher Britton (Dr Doom), Michael Adamthwaite (Justin Hammer), Ron Halder (Magneto), Andrew Francis (Hawkeye), Fred Henderson (Howard Stark), Peter Kelamis (Whiplash), Kristie Marsden (Madame Masque), Mark Oliver (Crimson Dynamo), David Orth (Blizzard), Tabitha St Germain (Maria Hill), Venus Terzo (Jean Grey), Lee Tockar (MODOK)

**Prod:** Joshua Fine, Cort Lane, Cédric Pilot, Romain van Liemt; **Music:** Guy Michelmore; **Exec.Prod:** Aton Soumache, Alexis Vonars, Dimitri Rassam, Stan Lee, Eric S Rollman, Tapaas Chakravarti, Lilian Eche, Steve Christian, Trevor Drinkwater, Stephen K. Bannon; **Developed by** Ciro Nieli, Joshua Fine, Christopher Yost; **Isle Of Man Film/Centre National de al Cinématrographie**; 22

**Notes:** Released after the success of **Iron Man**, but designed to capture the attention of children.

**Ratings:** *IMDB:* 67%

**Review:** It's weird to think that after watching **Iron Man**, a studio exec said "that's what will get us ratings! A show just like that movie…except all about kids." Having said that, the series is not really awful, though the animation style does take some getting used to. The other hurdle to get past is the fact this features teenage versions of the characters, which does stretch credulity in order to bring in other characters. It's not the worst way to while away the hours, but there are better things to be doing, all the same.

## THE SUPER HERO SQUAD SHOW
### (September, 2009 – October, 2011/Cartoon Network)

The Infinity Sword is shattered, and Dr Doom forms the Lethal Legion to reforge it, while Iron Man forms the Super Hero Squad to stop him. Each side sits in a city, separated by a giant wall and the Squad works hard to stop the Legion, little knowing that Galactus is approaching. And behind him, the only one who can wield the Infinity Sword; Thanos.

**Regular Cast:** *Charlie Adler (Dr Doom), Alimi Ballard (Falcon), Steven Blum (Wolverine), Dave Boat (Thor), Jim Cummings (Human Torch), Grey DeLisle (Ms Marvel), Mikey Kelley (Iron Fist), Tom Kenny (Iron Man),* Stan Lee (Mayor), *Tara Strong (Scarlet Witch), Travis Willingham (Hulk)*

**Episodes:**

*1.1 And Lo...A Pilot Shall Come!*
*1.2 To Err Is Superhuman!*
*1.3 This Silver, This Surfer!*
*1.4 Hulk Talk Smack!*
*1.5 Enter: Dormammu!*
*1.6 A Brat Walks Among Us!*
*1.7 From The Atom...It Rises!*
*1.8 Night in the Sanctorum!*
*1.9 This Forest Green!*
*1.10 O, Captain, My Captain!*
*1.11 If This Be My Thanos!*
*1.12 Deadly is the Black Widow's Bite!*
*1.13 Tremble At The Might Of... MODOK!*
*1.14 Mental Organism Designed Only For Kisses!*
*1.15 Invader from the Dark Dimension!*
*1.16 Tales of Suspense!*
*1.17 Strange From a Savage Land!*
*1.18 Mysterious Mayhem at Mutant High!*
*1.19 Election of Evil!*
*1.20 Oh Brother!*
*1.21 Hexed, Vexed And Perplexed!*
*1.22 The Ice Melt Cometh!*
*1.23 Wrath of the Red Skull!*
*1.24 Mother of Doom!*
*1.25 Last Exit Before Doomsday!*
*1.26 This Al Dente Earth!*

*2.1 Another Order Of Evil, Part 1!*
*2.2 Another Order Of Evil, Part 2!*
*2.3 World War Witch!*
*2.4 Villainy Redux Syndrome!*
*2.5 Support Your Local Sky-Father!*
*2.6 Whom Continuity Would Destroy!*
*2.7 Double Negation at The World's End!*
*2.8 Alienating With the Surfer!*
*2.9 Blind Rage Knows No Color!*
*2.10 Lo, How The Mighty Hath Abdicated!*
*2.11 So Pretty When They Explode!*
*2.12 Too Many Wolverines!*
*2.13 Pedicure and Facial of Doom!*
*2.14 Fate of Destiny!*
*2.15 The Ballard of Beta Ray Bill! (Six Against Infinity, Part 1)*
*2.16 Days, Nights and Weekends of Future Past! (Six Against Infinity, Part 2)*
*2.17 This Man-Thing, This Monster! (Six Against Infinity, Part 3)*
*2.18 The Devil Dinosaur You Say! (Six Against Infinity, Part 4)*
*2.19 Planet Hulk! (Six Against Infinity, Part 5)*
*2.20 1602! (Six Against Infinity, Part 6)*
*2.21 Brouhaha at the World's Bottom!*

*2.22 Missing: Impossible!*  *2.25 When Strikes the Surfer!*
*2.23 Revenge of the Baby Sat!*  *2.26 The Final Battle! ('Nuff Said!)*
*2.24 Soul Stone Picnic!*

**Notable Guest Cast:** *Shawn Ashmore (Iceman), John Barrowman (Stranger), Ty Burrell (Captain Marvel), Levar Burton (War Machine), Taye Diggs (Black Panther), Michael Dorn (Ronan), Robert Englund (Dormammu), Jonathan Frakes (High Evolutionary), Greg Grunberg (Ant Man), Mark Hamill (Red Skull), Lena Headey (Black Widow), Tricia Helfer (Sif), Wayne Knight (Egghead), Jane Lynch (Nebula), James Marsters (Mr Fantastic), Jennifer Morrison (Wasp), Jim Parsons (Nightmare), Adrian Pasdar (Hawkeye), Katee Sackoff (She-Hulk), Kevin Sorbo (Ka-Zar), Ray Stevenson (Punisher), George Takei (Galactus), Michelle Trachtenberg (Valkyrie), Adam West (Nighthawk)*

**Prod:** Dana C Booton; **Music:** Guy Michelmore; **Exec.Prod:** Eric S Rollman, Alan Fine, Simon Phillips, Joe Quesada; **Marvel Animation/Film Roman Productions**; 22

**Notes:** Designed as a series for the entire family, the animation style was exaggerated and humour was the most important element.

**Ratings:** *IMDB:* 62%

**Review:** Very much a kids' series – though that's not in itself a bad thing – but contrary to intentions, it doesn't hold an awful lot for adults. There are a few fun gags, but mostly this is very much for the children. It satisfies its target audience perfectly, has an astonishing guest cast, and it's difficult to dislike.

## *BLACK PANTHER*
### *(January, 2010/ABC)*

As T'Challa becomes the new Black Panther, the American government learns more about the secret nation of Wakanda and its protector, the Black Panther. T'Challa investigates his father's death, and learns that an assassin named Klaw is assembling a group to attack his country.

**Regular Cast:** *Djimon Hounsou (Black Panther/T'Challa),* Stan Lee (General Wallace), *Kerry Washington (Princess Shuri),* Alfre Woodard (Queen Mother), Carl Lumbly (Uncle S'Yan), *Jill Scott (Storm), Stephen Stanton (Klaw)*

**Episodes:**

*1.1 Pilot*
*1.2 Black Panther*
*1.3 Revenge of the Evil*
*1.4 Death of Father*
*1.5 Black Panther Vs Juggernaut and Black Knight*
*1.6 To the End*

**Notable Guest Cast:** *Jonathan Adams (T'Chaka), J B Blanc (Black Knight), Adrian Pasdar (Captain America), Kevin Michael Richardson (Wolverine), Rick D Wasserman (Radioactive Man), Peter Lurie (Juggernaut), Nolan North (Cyclops)*

**Dir:** Mark Brooks; **Prod:** Eric S Rollman; **Exec.Prod:** Sidney Clifton, Keith Fay, Aaron Parry, Reginald Hudlin, Cort Lane, Chris Prynoski, Shannon Barrett Prynoski, Denys Cowan; **Marvel Knights/BET Networks/Viacom**; 20

**Notes:** A short series developed by BET, which essentially was a motion comic. Curiously this was shown first in Australia, over a year before BET finally broadcast it. Originally Juggernaut was to have been the Rhino, which would better explain why he fights a black rhinoceros.

**Ratings:** *IMDB:* 74%

**Review:** A laudable attempt to do the series, but truthfully, motion comics can be hard to sit through, and one that is as talky as this one is particularly painful. With so little actually happening, the twenty minute episodes can seem to achieve nothing. Not really worth it.

## *THE AVENGERS: EARTH'S MIGHTIEST HEROES*
### *(September, 2010 – May, 2013/Disney XD)*

Four of SHIELD's prisons are busted open and seventy-five super powered criminals escape. Tony Stark assembles a team including Captain America, Ant-Man, Hulk, Thor and Wasp to recapture the criminals, little knowing that Thor's brother Loki was behind the original breakout.

**Regular Cast:** *Brian Bloom (Captain America), Chris Cox (Hawkeye), Jennifer Hale (Ms Marvel), Peter Jessop (Vision), Phil LaMarr (JARVIS), Eric Loomis*

(Iron Man), James C Mathis III (Black Panther), Colleen O'Shaughnessey (Wasp), Fred Tatasciore (Hulk), Rick D Wasserman (Thor), Wally Wingert (Ant-Man)

**Episodes:**

*0.1 Iron Man Is Born!*
*0.2 The Coming of the Hulk*
*0.3 The Man in the Ant Hill*
*0.4 HYDRA Lives*
*0.5 Thor the Mighty*
*0.6 Behold, The Mandroids!*
*0.7 Hulk Vs the World*
*0.8 The Siege of Asgard*
*0.9 Nick Fury, Agent of SHIELD*
*0.10 This Monster, This Hero*
*0.11 My Brother, My Enemy*
*0.12 The Isle of Silence*
*0.13 Enter the Whirlwind*
*0.14 Meet Captain America*
*0.15 The Red Skull Strikes!*
*0.16 If This Be Doomsday!*
*0.17 Welcome to Wakanda*
*0.18 Lo, There Shall Come A Conqueror*
*0.19 Beware the Widow's Bite!*
*0.20 The Big House*
*1.1 Breakout Part 1*
*1.2 Breakout Part 2*
*1.3 Iron Man Is Born*
*1.4 Thor the Mighty*
*1.5 Hulk Vs the World*
*1.6 Meet Captain America*
*1.7 The Man in the Ant Hill*
*1.8 Some Assembly Required*
*1.9 Living Legend*
*1.10 Everything Is Wonderful*
*1.11 Panther's Quest*
*1.12 Gamma World Part 1*

*1.13 Gamma World Part 2*
*1.14 Masters of Evil*
*1.15 459*
*1.16 Widow's Sting*
*1.17 The Man Who Stole Tomorrow*
*1.18 Come the Conqueror*
*1.19 The Kang Dynasty*
*1.20 The Casket of Ancient Winters*
*1.21 Hail, HYDRA!*
*1.22 Ultron-5*
*1.23 The Ultron Imperative*
*1.24 This Hostage Earth*
*1.25 The Fall of Asgard*
*1.26 A Day Unlike Any Other*
*2.1 The Private War of Doctor Doom*
*2.2 Alone Against AIM*
*2.3 Acts of Vengeance*
*2.4 Welcome to the Kree Empire*
*2.5 To Steal an Ant-Man*
*2.6 Michael Korvac*
*2.7 Who Do You Trust?*
*2.8 The Ballad of Beta Ray Bill*
*2.9 Nightmare in Red*
*2.10 Prisoner Of War*
*2.11 Infiltration*
*2.12 Secret Invasion*
*2.13 Along Came A Spider...*
*2.14 Behold...The Vision!*
*2.15 Powerless*
*2.16 Assault On 42*
*2.17 Ultron Unlimited*
*2.18 Yellowjacket*
*2.19 Emperor Stark*

**Notable Guest Cast:** *Dee Bradley Baker (Mr Fantastic), Drake Bell (Spider-Man), Steve Blum (Wolverine), Clancy Brown (Odin), Lacey Chabert (Daisy Johnson), Jeffrey Combs (The Leader), Elizabeth Daily (Mockingbird), Alex Désert (Nick Fury), Robin Atkin Downes (Baron Zemo), Christopher B Duncan (Luke Cage), Keith Ferguson (Gen Ross), Crispin Freeman (Scott Lang), Mark Hamill (Ulysses Klaw), Lance Henriksen (Grim Reaper), Lex Lang (Dr Doom), Lorin Lester (Iron Fist), Gabriel Mann (Bruce Banner), Vanessa Marshall (Black Widow), Grahame McTavish (Loki), Dawn Olivieri (Pepper Potts), Lance Reddick (Falcon), Bumper Robinson (War Machine), Dwight Schultz (Technovore), J K Simmons (J Jonah Jameson), Brent Spiner (Purple Man), Erin Torpey (Invisible Woman), Jim Ward (Baron Von Strucker), Kari Wuhrer (Maria Hill)*

**Prod:** Joshua Fine, Dana C Booton; **Music:** Guy Michelmore; **Exec.Prod:** Simon Phillips, Stan Lee, Eric S Rollman, Joe Quesada; **Developed by** Ciro Nieli, Joshua Fine, Christopher Yost; **Marvel Animation/Film Roman Productions**; 22

**Notes:** A micro season of twenty episodes was broadcast before the series was broadcast proper. The episodes were then edited together to make episodes three – seven of the first season. The first season dealt with the core Avengers battling Loki and his freeing of a large number of super criminals. The second season revolved around the expanded team battling initially the Skrulls and then the Kree. Both seasons saw storylines borrow from the comics, including, notably, Secret Invasion. The release of **The Avengers** changed the direction that Marvel Animation wanted to head, and this series was cancelled. Although initially hinted that **Avengers Assemble** would be a continuation of **Earth's Mightiest Heroes**, It clearly wasn't.

**Ratings:** *IMDB:* 84%

**Review:** Following the tradition of **Wolverine and the X-Men**, this series really brings the Avengers to life in a way that does the comics credit. Expanding to include a variety of different areas from the Marvel universe including the X-Men, the Fantastic Four and some street level vigilantes, this series feels more like a Marvel Universe animation than specifically an Avengers one. Clever, engaging and respectful of the source material, its cancellation is probably the one big negative of Joss Whedon's movie.

## MARVEL ANIME
### (October, 2010 – April, 2012)

In Japan, Tony Stark prepares for the launch of his Arc Station, and his new Dio armour. However, the launch is sabotaged by the Zodiac, and after his Dio armour is stolen, Stark sets out to battle the Zodiac and retrieve it. Logan tracks down his love, Mariko Yashida, only to discover that her father, crime lord Shingen, has taken her in readiness for an arranged marriage to better cement his position. When Hisako Ichiki disappears, Professor Xavier reassembles the remains of his X-Men to discover what happened to her. They find that a group calling themselves the U-Men are harvesting mutants for their organs. Blade seeks out the vampire who killed his mother and soon finds himself having to kill two of his friends when they are turned by Deacon Frost – the very man he is seeking.

**Regular Cast:** *Keiji Fujiwara, Adrian Pasdar (Tony Stark/Iron Man)[1.1-1.12], Hiroaki Hirata, Kyle Hebert (Ho Yinsen/Iron Man Dio)[1.1-1.12], Hiroe Oka, Cindy Robinson (Pepper Potts)[1.1-1.12], Rikiya Koyama, Milo Ventimiglia [2.1-2.12], Steven Blum [3.1-3.12] (Logan/Wolverine)[2.1-3.12], Fumiko Orikasa, Gwendoline Yeo (Mariko Yashida)[2.1-2.12], Hidekatsu Shibata, Fred Tatasciore (Shingen Yashida)[2.1-2.12], Romi Park, Kate Higgins (Yukio)[2.1-2.12], Toshiyuki Morikawa, Scott Porter (Scott Summers/Cyclops)[2.1-2.12], Katsunosuke Hori, Cam Clarke (Charles Xavier)[3.1-3.12], Yurkia Hino, Jennifer Hale (Jean Grey)[3.1-3.12], Aya Hisakawa, Danielle Nicolet (Ororo Munroe/Storm)[3.1-3.12], Haruhiko Joo, Travis Willingham (Jason Wyngarde/Mastermind)[3.1-3.12], Hideyuki Tanaka, Fred Tatasciore (Hank McCoy/Beast)[3.1-3.12], Karoi Yamagata, Ali Hillis (Emma Frost)[3.1-3.12], Yukari Tamura, Stephanie Sheh (Hisako Ichiki/Armor)[3.1-3.12], Akio Ohtsuka, Harold Perrineau (Eric Brooks/Blade)[4.1-4.12], Tsutomu Isobe, J B Blanc (Deacon Frost)[4.1-4.12]*

**Episodes:**

*1.1 Japan: Enter Iron Man*

*1.2 Going Nuclear*

*1.3 Reap the Whirlwind*

*1.4 A Twist of Memory, A Turn of the Mind*

*1.5 Outbreak*

*1.6 Technical Difficulties*

*1.7 At the Mercy of My Friends*

*1.8 Daughter of the Zodiac*

*1.9 A Duel of Iron*

*1.10 Casualties of War*

*1.11 The Beginning of the End*

**Prod:** Fuminomi Hara, Cort Lane, Taro Morishima; **Exec.Prod:** Dan Buckley, Alan Fine, Daisuke Gomi, Jeph Loeb, Simon Phillips, Eric S Rollman, Masao Takiyama, Yuuki Yoshida, Masao Maruyama; **Marvel Studios/Madhouse**

**Notes:** Billed alternately as a four series program, or four individual television programs, this is a Japanese anime production of Iron Man, Wolverine, X-Men and Blade. The dual cast represents both the Japanese speaking and English speaking sides. The series was given free rein to reinvent the Marvel universe as it saw fit.

**Ratings:** *IMDB:* 67%

## ULTIMATE SPIDER-MAN
### (April 2012 – January, 2017/Disney XD)

Spider-Man is recruited by SHIELD (and disturbingly discovers his new principal is a SHIELD agent), with a view to training him to be a true superhero. Joining forces with a team of other training heroes – Nova, White Tiger, Iron Fist and Power Man – Spider-Man and his team take on the villains of New York.

**Regular Cast:** *Drake Bell (Peter Parker/Spider-Man), Ogie Banks (Luke Cage/Power Man), Greg Cipes (Danny Rand/Iron Fist), Caitlyn Taylor Love (Ava Ayala/White Tiger), Logan Miller (Sam Alexander/Nova), Clark Gregg (Phil Coulson), Chi McBride (Nick Fury), J K Simmons (J Jonah Jameson), Tom Kenny (Dr Octopus), Matt Lanter (Harry Osborn), Steven Weber (Norman Osborn)*

## Episodes:

*1.1 Great Power (Part 1)*
*1.2 Great Responsibility (Part 2)*
*1.3 Doomed!*
*1.4 Venom*
*1.5 Flight of the Iron Spider*
*1.6 Why I Hate Gym*
*1.7 Exclusive*
*1.8 Back in Black*
*1.9 Field Trip*
*1.10 Freaky*
*1.11 Venomous*
*1.12 Me Time*
*1.13 Strange*
*1.14 Awesome*
*1.15 For Your Eye Only*
*1.16 Beetle Mania*
*1.17 Snow Day*
*1.18 Damage*
*1.19 Home Sick Hulk*
*1.20 Run Pig Run*
*1.21 I Am Spider-Man*
*1.22 The Iron Octopus*
*1.23 Not a Toy*
*1.24 Attack of the Beetle*
*1.25 Revealed (Part 1)*
*1.26 Rise of the Goblin (Part 2)*
*2.1 The Lizard*
*2.2 Electro*
*2.3 The Rhino*
*2.4 Kraven the Hunter*
*2.5 Hawkeye*

*2.6 The Sinister Six*
*2.7 Spidah-Man!*
*2.8 Carnage*
*2.9 House Arrest*
*2.10 The Man-Wolf*
*2.11 Swarm*
*2.12 Itsy Bitsy Spider-Man*
*2.13 Journey of the Iron Fist*
*2.14 The Incredible Spider-Hulk*
*2.15 Stan by Me*
*2.16 Ultimate Deadpool*
*2.17 Venom Bomb*
*2.18 Guardians of the Galaxy*
*2.19 The Parent Trap*
*2.20 Game Over*
*2.21 Blade (Part 1)*
*2.22 The Howling Commandos (Part 2)*
*2.23 Second Chance Hero*
*2.24 Sandman Returns*
*2.25 Return of the Sinister Six (Part 1)*
*2.26 Ultimate (Part 2)*
*3.1 The Avenging Spider-Man (Part 1)*
*3.2 The Avenging Spider-Man (Part 2)*
*3.3 Agent Venom*
*3.4 Cloak and Dagger*
*3.5 The Next Iron Spider*
*3.6 The Vulture*

*3.7 The Savage Spider-Man*

*3.8 New Warriors*

*3.9 The Spider-Verse (Part 1)*

*3.10 The Spider-Verse (Part 2)*

*3.11 The Spider-Verse (Part 3)*

*3.12 The Spider-Verse (Part 4)*

*3.13 The Return of the Guardians of the Galaxy*

*3.14 SHIELD Academy*

*3.15 Rampaging Rhino*

*3.16 Ant-Man*

*3.17 Burrito Run*

*3.18 Inhumanity*

*3.19 Attack of the Synthezoids (Part 1)*

*3.20 The Revenge of Arnim Zola (Part 2)*

*3.21 Halloween Night at the Museum*

*3.22 Nightmare on Christmas*

*3.23 Contest of Champions (Part 1)*

*3.24 Contest of Champions (Part 2)*

*3.25 Contest of Champions (Part 3)*

*3.26 Contest of Champions (Part 4)*

*4.1 Hydra Attacks (Part 1)*

*4.2 Hydra Attacks (Part 2)*

*4.3 Miles from Home*

*4.4 Iron Vulture*

*4.5 Lizards*

*4.6 Double Agent Venom*

*4.7 Beached*

*4.8 Anti-Venom*

*4.9 Force of Nature*

*4.10 The New Sinister 6 (Part 1)*

*4.11 The New Sinister 6 (Part 2)*

*4.12 Agent Web*

*4.13 The Symbiote Saga (Part 1)*

*4.14 The Symbiote Saga (Part 2)*

*4.15 The Symbiote Sage (Part 3)*

*4.16 Return to the Spider-Verse (Part 1)*

*4.17 Return to the Spider-Verse (Part 2)*

*4.18 Return to the Spider-Verse (Part 3)*

*4.19 Return to the Spider-Verse (Part 4)*

*4.20 Strange Little Halloween*

*4.21 The Spider Slayers (Part 1)*

*4.22 The Spider Slayers (Part 2)*

*4.23 The Spider Slayers (Part 3)*

*4.24 The Moon Knight Before Christmas*

*4.25 Graduation Day (Part 1)*

*4.26 Graduation Day (Part 2)*

**Notable Guest Cast:** *Donald Glover (Miles Morales/Spider-Man), Freddy Rodriguez (Miguel O'Hara/Spider-Man), Milo Ventimiglia (Spider-Man Noir), Troy Baker (Hawkeye), Laura Bailey (Black Widow), David Kaye (JARVIS), Adrian Pasdar (Iron Man), Bumper Robinson (Falcon), Roger Craig Smith (Captain America), Fred Tatasciore (Hulk), Travis Willingham (Thor), Chris Box (Peter Quill), Trevor Devall (Rocket), Michael Clarke Duncan (Groot), Nika Futterman (Gamora), David Sobolov (Drax), Steven Blum (Wolverine), Will Friedle (Deadpool), Dave Boat (The Thing), Jack Coleman (Dr Strange), Terry Crews (Blade), Ashley Eckstein (Dagger), Phil LaMarr (Cloak), Iain De Caestecker (Leo Fitz), Elizabeth Henstridge (Jemma Simmons), Mark Hamill (Nightmare), Misty Lee (Aunt May), Stan Lee (Stan the Janitor)*

**Prod:** Brian Michael Bendis, Paul Dini; **Music:** Kevin Manthei; **Exec.Prod:** Alan Fine, Joe Quesada, Dan Buckley, Jeph Loeb, Steven T Seagle, Duncan Rouleau, Joe Kelly, Joe Casey; **Marvel Animation/Film Roman**; 22

**Notes:** *Ultimate Spider-Man* was the longest running Marvel television show ever made. The third season had a subtitle – Web Warriors. Although the season three episode "Halloween Night at the Museum" is officially the 21st episode of the season, it was broadcast well before that; skipped forward by almost a year to be broadcast on Halloween, after episode seven. Similarly "Nightmare on Christmas" was broadcast between episodes eight and nine, despite being the 22nd official episode.

**Ratings:** *IMDB:* 71%

**Review:** Credit has to be due to a series that can be quite as surreal as this one. The decision to have Spider-Man break the fourth wall and talk with the audience, as well as interact with chibi versions of the characters, is a brave one, but does actually pay off. Additionally, given that it takes time to use so many elements of the comics – including a Spider-verse arc – it's hard not to enjoy this series. A worthwhile cartoon for adults and children alike.

## *AVENGERS ASSEMBLE*
### *(May, 2013 – February, 2019/Disney XD)*
Because of the super soldier serum he used, Red Skull is slowly dying, and he recruits MODOK to help him capture Captain America in the hope of changing bodies. Though his plans don't come to fruition, thanks to the Avengers, Skull is able to get hold of an Iron Man armour, which allows him to gather more allies in his quest. Unknown to the Avengers, however, both Thanos and Ultron lurk just around the corner…

**Regular Cast:** *Laura Bailey (Black Widow), Troy Baker (Hawkeye), Adrian Pasdar (Iron Man), Bumper Robinson (Falcon), Roger Craig Smith (Captain America), Fred Tatasciore (Hulk), Travis Willingham (Thor)*

**Episodes:**

*1.1 The Avengers Protocol (Part 1)*    *1.4 The Serpent of Doom*
*1.2 The Avengers Protocol (Part 2)*    *1.5 Blood Feud*
*1.3 Ghost of a Chance*    *1.6 Super-Adaptoid*

4.3 *The Sleeper Awakens*
4.4 *Prison Break*
4.5 *The Incredible Herc*
4.6 *Show Your Work*
4.7 *Sneakers*
4.8 *Why I Hate Halloween*
4.9 *The Once and Future Kang*
4.10 *Dimension Z*
4.11 *The Most Dangerous Hunt*
4.12 *Under the Spell of the Enchantress*
4.13 *The Return*
4.14 *New Year's Resolution*
4.15 *The Eye of Agamotto (Part 1)*
4.16 *The Eye of Agamotto (Part 2)*
4.17 *Beyond*
4.18 *Underworld*
4.19 *The Immortal Weapon*
4.20 *The Vibranium Coast*
4.21 *Weirdworld*
4.22 *Westland*
4.23 *The Citadel*
4.24 *The Wastelands*
4.25 *All Things Must End*

5.1 *Shadow of Atlantis (Part 1)*
5.2 *Shadow of Atlantis (Part 2)*
5.3 *Into the Deep*
5.4 *The Panther and the Wolf*
5.5 *The Zemo Sanction*
5.6 *Mists of Attilan*
5.7 *T'Challa Royale*
5.8 *The Night Has Wings*
5.9 *Mask of the Panther*
5.10 *The Good Son*
5.11 *The Lost Temple*
5.12 *Descent of the Shadow*
5.13 *The Last Avenger*
5.14 *The Vibranium Curtain (Part 1)*
5.15 *The Vibranium Curtain (Part 2)*
5.16 *T'Chanda*
5.17 *Yemandi*
5.18 *Bashenga*
5.19 *King Breaker (Part 1)*
5.20 *King Breaker (Part 2)*
5.21 *Widowmaker*
5.22 *Atlantis Attacks*
5.23 *House of M*

**Notable Guest Cast:** *Charlie Adler (MODOK), Drake Bell (Spider-Man), Stephen Collins (Howard Stark), Chris Cox & Will Friedle (Star-Lord), Nika Futterman (Gamora), Grant George (Ant-Man), Seth Green & Trevor Devall (Rocket Raccoon), David Kaye (JARVIS), Maurice LaMarche (Dr Doom), Chi McBride (Nick Fury), Kevin Michael Richardson (Groot), Dwight Schultz (Attuma), J K Simmons (J Jonah Jameson), David Sobolov (Drax the Destroyer), Frank Welker (Odin), Hayley Atwell ) Peggy Carter), Wynn Everett (Whitney Frost)*

**Prod:** Joe Casey, Joe Kelly, Duncan Rouleau, Steven T Seagle; **Music:** Michael McCuistion, Lolita Ritmanis, Kristopher Carter; **Exec.Prod:** Alan Fine, Joe Quesada, Dan Buckley, Jeph Loeb, Man of Action; **Marvel Animation/ Man Of Action Studios**; 22

**Notes:** This series replaced *The Avengers: Earth's Mightiest Heroes*, but was effectively a reboot to bring the series in line with the recently released live action movie. As such the Avengers line up is mostly that of the movie, and the storylines are new. The first season revolves around the Red Skull assembling a Cabal to fight the Avengers. The second concerns Thanos' quest to gain the Infinity Stones. The third season is subtitled *Ultron Revolution* and is set after the new movie. In spite of this, the series is not part of the MCU, but does seem to be part of the same universe with *Ultimate Spider-Man* and *Hulk and the Agents of SMASH*. Actors from all three series cross over with each other, though storylines never do. Additionally, the actors from this series appeared in *Phineas & Ferb* as their Marvel characters. Stan Lee recorded a voice for 5.16 not long before his death; 5.23 would then be dedicated to him.

**Ratings:** *IMDB:* 70%

**Review:** An entertaining television show, but sadly, due to its desire to be more in line with the movies, it seems scared to have the same depth and scope of its predecessor. Conversely, the series never does feel like the movies though, particularly by never really having Bruce Banner appear. By the third series, the program picks up, and the introduction of other Avengers starts to give a more independent feel to the series. Ultimately, light and easily devourable; good for kids.

## HULK AND THE AGENTS OF SMASH
### (August, 2013 – June, 2015/Disney XD)

Against his better judgement, Hulk joins forces with Red Hulk, Skaar, She-Hulk and A-Bomb to take on a variety of super powered menaces that threaten the Earth, in the hope of proving Hulk's good character.

**Regular Cast:** *Fred Tatasciore (Bruce Banner/Hulk), Seth Green (Rick Jones/ A-Bomb), Eliza Dushku (Jennifer Walters/She-Hulk), Clancy Brown (Thunderbolt Ross/Red Hulk), Benjamin Diskin (Skaar)*

**Episodes:**

*1.1 Doorway to Destruction (Part 1)*   *1.6 Savage Land*
*1.2 Doorway to Destruction (Part 2)*   *1.7 The Incredible Shrinking Hulks*
*1.3 Hulk-Busted*   *1.8 Hulks on Ice*
*1.4 The Collector*   *1.9 Of Moles and Men*
*1.5 All About the Ego*   *1.10 Wendigo Apocalypse*

**Notable Guest Cast:** *James Arnold Taylor (The Leader), J K Simmons (J Jonah Jameson), Robin Atkin Downes (Abomination), Kevin Michael Richardson (Ego The Living Planet), Stan Lee (Mayor Stan), Adrian Pasdar (Iron Man), Travis Willingham (Thor), Drake Bell (Spider-Man), John DiMaggio (Galactus), Brent Spiner (Silver Surfer), Robert Englund (Pluto), Jack Coleman (Dr Strange), Terry Crews (Blade), Frank Welker (Odin)*

**Super.Prod:** Cort Lane, Henry Gilroy, Mitch Schauer, Todd Casey, Harrison Wilcox, Stephen Wacker; **Music:** Guy Michelmore; **Exec.Prod:** Alan Fine, Joe Quesada, Dan Buckley, Jeph Loeb; **Ed:** Adam Redding; **Marvel Animation/ Film Roman Productions**; 22

**Notes:** Another in the *Ultimate Spider-Man/Avengers Assemble* universe, though the continuity doesn't quite match up – Hulk living with the Avengers and working with them full time in *Avengers Assemble*, doesn't really tally with the fact he is living full time with SMASH and working with them full time.

The cast for this series is used in the other two series, as well as other series where Marvel characters turn up.

**Ratings:** *IMDB:* 64%

**Review:** Despite a standout cast, this is easily the weakest of the three animated Marvel Universe series. The series continues with the habit of barely using Bruce Banner, and hardly dealing with the conflict between Banner and the Hulk. Although light like the other series, it doesn't carry the humour as well as ***Ultimate Spider-Man***, and lacks even the minimal depth of ***Avengers Assemble***. An easy series to pass on.

## *LEGO MARVEL SUPER HEROES: MAXIMUM OVERLOAD*
### *(11 November, 2013/Netflix)*

With his enchanted snowballs allowing him to overload villains, Loki decides to wreak havoc on the super heroes of New York.

**Regular Cast:** *Laura Bailey (Black Widow), Dee Bradley Baker (Venom, Chitauri), Troy Baker (Loki), Drake Bell (Spider-Man), Steve Blum (Wolverine), Greg Cipes (Iron Fist), Grey DeLisle (Pepper Potts), Barry Dennen (Mandarin), Robin Atkin Downes (Abomination), Tom Kenny (Doctor Octopus), Stan Lee (Hot Dog Vendor), Tony Matthews (Jogger), Chi McBride (Nick Fury, Red Skull), Adrian Pasdar (Iron Man), Bumper Robinson (Falcon), J K Simmons (J Jonah Jameson), Roger Craig Smith (Captain America), Fred Tatasciore (Hulk), Travis Willingham (Thor)*

### Episodes:

**1.1 A Faceful Of Danger!**
**1.2 Slaughter on the 23rd Floor!**
**1.3 Obnoxious Mandarin!**
**1.4 Operation Doofus Drop!**
**1.5 Assault, Off-Asgard!**

**Notable Guest Cast:** *Jonathan Adams (Ronan the Accuser), Jeff Bennett (Rhomann Day), John DiMaggio (Lunatik), Dave Fennoy (Korath), Tom Kenny (The Collector), Isaac Singleton Jr (Thanos) , Jason Spisak (Grandmaster), Tara Strong (Irani Rael), Cree Summer (Nebula), James Arnold Taylor (Yondu)*

**Dir:** Greg Richardson; **Writer:** Matt Wayne; **Prod:** Tony Matthews; **Music:** Asher Lenz & Stephen Skratt; **Exec.Prod:** Jill Wilfert, Jason Cosler, Alan Fine, Dan Buckley, Joe Quesada, Jeph Loeb, Kallan Kagan, Marianne Culbert; **Ed:** Jeremy Montgomery; **Marvel Entertainment/Lego**

**Notes:** A five-part miniseries that was ultimately re-released as a 22 minute episode, this saw Marvel and Lego teaming up to create something akin to the style of the Lego Marvel video games.

**Ratings:** *IMDB:* 67%

**Review:** Heaps of fun if you like that sort of thing, this is Lego and Marvel not taking themselves remotely seriously, but still delivering an entertaining storyline. Children will adore it, and if adults are prepared to let go of their maturity, they will as well.

## *DISK WARS: THE AVENGERS* (ディスク・ウォーズ：アベンジャーズ, )
### *(April, 2014 – March, 2015/TXN)*

Tony Stark and Nozomu Akatsuki develop the Digitial Identity Securement Kit – or DISK – which is an effective method of detaining super powered individuals. However, Loki is able to turn the tables, trapping the Avengers in the DISKs instead, with only Nozomu's two sons and their three friends able to release the heroes from their DISKs for short periods of time. With this limitation, the Avengers set out to find the DISKs that contain the biocodes for the villains and recapture them.

**Regular Cast:** Mitsuki Saiga (Akira Akatsuki), Yuuichi Iguchi (Hikaru Akatsuki), Yayoi Sugaya (Edward Grant), Yuusuke Kuwahata (Chris Taylor), Naomi Oozora (Jessica Shannon), *Eiji Hanawa (Iron Man), Kazuhiro Nakaya (Captain America), Yasuyuki Kase (Thor), Kenichirou Matsuda (The Hulk), Kaori Mizuhashi (The Wasp), Tadashi Mutou (Loki)*

**Episodes:**

| | |
|---|---|
| **1.1 The Mightiest Of Heroes** | **1.7 Hulk Versus Captain** |
| **1.2 Heroes Annihilates?!** | **1.8 Avengers Assemble** |
| **1.3 Iron Man, Reborn!** | **1.9 Spider-Man Is Missing** |
| **1.4 Re-enforcement Hawkeye!** | **1.10 Showdown! The Silver Donned** |
| **1.5 Mighty Thor Descends** | **Samurai!** |
| **1.6 Hulk Runs Wild** | **1.11 A Gift from Tony** |

**Notable Guest Cast:** *Shinji Kawada (Spider-Man)*

**Prod:** King Ryuu; **Music:** T M Revolution; **Toei Animation/Walt Disney Television International Japan**

**Notes:** This series is essentially a Marvel version of something like Pokemon, with the Marvel characters being captured in devices called Digital Identity Securement Kits (DISK) which a group of children possess and can release the Marvel heroes from.

**Ratings:** *IMDB:* 47%

**Review:** Pretty much as insane as you'd expect from this kind of television show, it really does feel like a standard Japanese anime with Marvel characters shoe-horned it. Despite the fact it has to work within the confines of the merchandising requirements, the program is still quite fun, if you enjoy that sort of thing.


## GUARDIANS OF THE GALAXY
### (July, 2015 – June, 2019/Disney XD)

Peter Quill and his team rescue Yondu, who wants them to help him recover a Spartax cube from Korath. The artefact is actually a map to the cosmic seed, but it requires Pandorian crystals to power it. The search is on for Pandorian crystals, as Thanos hunts them down.

**Regular Cast:** *Will Friedle (Peter Quill/Star-Lord), Vanessa Marshall (Gamora), Trevor Devall (Rocket), Kevin Michael Richardson (Groot), David Sobolov (Drax)*

**Episodes:**

*1A.1 Star-Lord*
*1A.2 Groot*
*1A.3 Rocket Racoon*
*1A.4 Drax*
*1A.5 Gamora*
*1.1 Road to Knowhere (Part 1)*
*1.2 Knowhere to Run (Part 2)*
*1.3 One in a Million You*
*1.4 Take the Milano and Run*
*1.5 Can't Fight This Seedling*
*1.6 Undercover Angle*
*1.7 The Backstabbers*
*1.8 Hitchin' a Ride*
*1.9 We Are Family*
*1.10 Bad Moon Rising*
*1.11 Space Cowboy*
*1.12 Crystal Blue Persuasion*
*1.13 Stuck In the Metal with You*
*1.14 Don't Stop Believin'*
*1.15 Accidents Will Happen*

*1.16 We Are the World Tree*
*1.17 Come and Gut Your Love*
*1.18 Asgard War, Part 1: Lightin' Strikes*
*1.19 Asgard War, Part 2: Rescue Me*
*1.20 Fox on the Run*
*1.21 Inhuman Touch*
*1.22 Welcome Back*
*1.23 I've Been Searching So Long*
*1.24 I Feel the Earth Move*
*1.25 Won't Get Fooled Again*
*1.26 Jingle Bell Rock*
*2A.1 Pick Up the Pieces*
*2A.2 Star-Lord Vs MODOK*
*2A.3 Drax Attacks!*
*2A.4 Rocket! Groot! Man-Thing!*
*2A.5 Gamora Strikes!*
*2A.6 Guardians Reunited!*
*2.1 Stayin' Alive*
*2.2 Evolution Rock*

**Notable Guest Cast:** Jonathan Adams (Ronan the Accuser), Jeff Bennett (Rhomann Day), John DiMaggio (Lunatik), Dave Fennoy (Korath), Tom Kenny (The Collector), Isaac Singleton Jr (Thanos) , Jason Spisak (Grandmaster), Tara Strong (Irani Rael), Cree Summer (Nebula), James Arnold Taylor (Yondu)

**Super.Prod:** Harrison Wilcox, Marty Isenberg; **Music:** Michael Tavera; **Exec.Prod:** Alan Fine, Joe Quesada, Dan Buckley, Jeph Loeb; **Marvel Animation**; 22

**Notes:** After the huge success of the **Guardians of the Galaxy** film, when

Marvel Animation came to creating a fourth series to add to the animated universe, this became the obvious choice – not least because they'd already tested well in *Ultimate Spider-Man* and *Avengers Assemble*. The voice cast from those two series – for the most part – transferred to this series, and the series sort of continues on from the movie, utilising songs from the 1970's. 1A. 1 – 1A.5 were five short animations showing some of the backstory of the characters.

**Ratings:** *IMDB:* 77%

**Review:** Better than the *Hulk* series, *Guardians* still doesn't seem quite as interesting as the other two series in the universe, and strangely enough the animation on this one seems a lot more rudimentary, which really doesn't work in its favour. The series has grown stronger since moving away from the movie that inspired it, and the storylines centring on Quill's father are an intriguing take on the story.

## *FUTURE AVENGERS*
### *(July, 2017 – October, 2018 /Dlife)*

Makoto has had his genes manipulated by Hydra giving him special abilities. Alongside Aki and Chloe, he joins with the Avengers to learn how to become a superhero.

**Regular Cast:** *Hideaki Hanawa/Mick Wingert (Tony Stark/Iron Man), Kazuhiro Nakatani/Roger Craig Smith (Steve Rogers/Captain America), Yasuyuki Kase/ Patrick Seitz (Thor), Kenichiro Matsua/Fred Tatasciore (Hulk), Mizuhashi Kaori/Kari Wahlgren (Wasp),* Aki Kanada/Max Mittelman (Makoto), Atsushi Tamaru/Xander Morbus (Akiyadi), Juri Kimura/Dina Sherman (Chloe)

**Episodes:**

*1.1 Aim for the Avengers*
*1.2 Red Skull of Despair*
*1.3 Red Skull Fight*
*1.4 Makoto's Rebellion*
*1.5 The Unexplored City of K'un-Lun*
*1.6 Hero's Qualification*
*1.7 I Came! Deadpool!*

*1.8 Cap's Distant Past*
*1.9 Winter Soldier's Identity*
*1.10 Black Panther Protecting the Kingdom*
*1.11 Mysterious Hulk Runaway*
*1.12 Hero Control Law Passed*
*1.13 Green Monster Vs Green Giant*
*1.14 Super Fast Flight Which Can't*

*be Stopped*

*1.15 Another Transformation Girl*

*1.16 Winter Soldier Rescue Battle*

*1.17 Find Hydra's Collaborators*

*1.18 Deadpool Again*

*1.19 Do you Believe in Loki!?*

*1.20 Bonds of "Brothers"*

*1.21 The Worst Conquistador*

*1.22 Absolute Fortress*

*1.23 Makoto Taking Time*

*1.24 Breakthrough Breaking Strategy!*

*1.25 Final Battle of Destiny*

*1.26 Future Avengers*

*2.1 Visco's Revolt!?*

*2.2 Sneak in! Dark Auction*

*2.3 With Nyan? Hawkeye*

*2.4 A Boy Drawing a Monster Picture*

*2.5 Monster Battle*

*2.6 Mystery, Mist and Ability*

*2.7 Bad Organisation AIM Research*

*2.8 Replication!? Power of Justice*

*2.9 Inhumans' Entrance*

*2.10 Lunar City Atyran*

*2.11 Showdown! Black Bolt*

*2.12 Advance! Maximus Fleet*

*2.13 At the End of the Battle*

**Dir:** Yuzo Sato; **Music:** King Ryu; **Exec.Prod:** Alan Fine, Joe Quesada, Dan Buckley, Stan Lee, C B Cebulski, Cort Lane; **Walt Disney Japan/Madhouse**; 25

**Notes:** An obscure one this – another Japanese Marvel animated series, that has only recently been given an English dub release. (The second actors in the cast listing are the English voice actors). The theme is heavily influenced by Alan Silvestri's **Avengers'** theme. It was finally shown on Disney+ in 2020, allowing people much easier access to the show. Although it is fairly typical anime in design, it does still look great. Though never formally cancelled, a third season was never commissioned.

**Ratings:** *IMDB:* 57%

### SPIDER-MAN
#### *(August, 2017 – October, 2020 /Disney XD)*

Peter Parker has recently gained the powers of a spider and has decided to become Spider-Man in order to fight crime. However, he's also recently been accepted into Horizon High, a school for genius scientists. Between this and keeping his friendships, Peter struggles to balance his life.

**Regular Cast:** *Robbie Daymond (Peter Parker/Spider-Man), Laura Bailey (Gwen Stacy), Nadji Jeter (Miles Morales), Melanie Minichino (Anya/Corazon/ Spider-Girl), Max Mittelman (Harry Osborn/Hobgoblin), Fred Tatasciore (Max Modell), Bob Joles (J Jonah Jameson)*

**Episodes:**

**Notable Guest Cast:** *John DiMaggio (Raymond Warren/The Jackal), Alistair Duncan (The Vulture), Scott Menville (Otto Octavius/Doctor Octopus), Josh Keaton (Norman Osborn), Ben Diskin (Flash Thompson), Bettina Kenney (Liz Allan), Nancy Linari (May Parker), Patton Oswalt (Ben Parker), Aubrey Joseph (Cloak), Olviia Holt (Dagger), Mark Hamill (Arnim Zola), Ernie Hudson (Robbie*

*Robertson),* Joe Quesada (Joe Q)

**Super.Prod:** Harrison Wilcox, Marty Isenberg; **Music:** Michael Tavera; **Exec.Prod:** Alan Fine, Joe Quesada, Dan Buckley, Jeph Loeb; **Marvel Animation**; 21

**Notes:** With the release of **Spider-Man: Homecoming** the decision was taken to end **Ultimate Spider-Man** and replace it with a new series, starting the story again, and giving Peter his homemade costume initially. The series was kicked off by six short episodes which showed how Peter gained his abilities. The third season had the season title *Maximum Venom*, and each episode was actually double length. On repeat broadcasts, these episodes have been split into two, doubling the number of episodes for the season.

**Ratings:** *IMDB:* 64%

**Review:** It's not the first time Marvel have opted to end a perfectly good animated series and replace it, and as before, there's nothing wrong with **Spider-Man** as such, but it lacks the charm of its predecessor and restarting the series from scratch just seems counter-intuitive (particularly in this day and age of programming). Ultimately, it's good…but it's not an improvement on **Ultimate Spider-Man**, which just leaves you wondering, why did they bother?

## MARVEL SUPER HERO ADVENTURES
### *(October, 2017 – September, 2020/Disney Junior)*
Spider-Man tells a variety of stories about how he has helped superheroes over the years fight against different villains.

**Regular Cast:** *Cole Howard (Peter Parker/Spider-Man), Brian Drummond (Iron Man/Groot), David Atar (Thor), James Blight (Hulk), Adrian Petriw (Ant-Man), Zac Siewert (Miles Morales), Aliza Vellani (Ms Marvel), Michael Daingerfield (Captain America), Jacqueline Samuda (Captain Marvel), Omari Newton (Black Panther), Emily Tennant (Ghost-Spider), Gigi Saul Guerrero (Spider-Girl), Derek Mack (Falcon), Marlie Collins (Wasp), Diana Kaarina (Nebula), Trevor Devall (Rocket), Toren Atkinson (Dr Strange), Elysia Rotaru (She-Hulk), Matt Cowlrick (Loki)*

**Episodes:**

1.1 One Big Mess
1.2 Uh, Oh, It's Magic!
1.3 That Drone Cat
1.4 Rock and Roll
1.5 Electric Youth
1.6 The Toys are Back in Town
1.7 Way Outer Space
1.8 Family Friendly
1.9 Stomp and Listen
1.10 I'm Positive!
2.1 Evil Mittens
2.2 Doctor Octopus's Garden
2.3 Mine!
2.4 Now That's Funny!
2.5 LISTEN
2.6 Sticky Mittens
2.7 Cloudy with a Chance of Smiles
2.8 That's Why They Make Chocolate & Vanilla
2.9 It's an Alien!
2.10 You Go High, I'll Go Low
S1 Be Like Spidey!
S2 Meet Captain Marvel!
S3 Meet Thor!
S4 Spidey Remix with Miles!
S5 Spidey Spyin' with Thor!
S6 Spidey Spyin' with Electro!
S7 Meet Ms Marvel!
S8 Spidey Spyin' with Miles!
S9 Spidey Remix: Villain Fails!
S10 Meet Black Panther!
3.1 Sorry Seems to be the Hardest Word
3.2 The Claws of Life
3.3 Not So Dry Cleaners
3.4 From Hulk to Eternity
3.5 Sticky Rain
3.6 Try It, You'll Like It!
3.7 Bend, Don't Break
3.8 That's What Friends Are For
3.9 Happy Birthday
3.10 Things That Go HaHa! In the Night
4.1 Charge Ahead
4.2 Aww...Do I Have To?
4.3 It's On Me
4.4 Promises Promises
4.5 Outside The Box
4.6 It's Clobberin' Time
4.7 It's Too Dam Hot
4.8 One For All And All 4 One
4.9 Building Bridges
4.10 Obstacle, Of Course

**Dir:** Svend Gregori; **Writer:** Eugene Son, David Kidd, Ron Burch; **Music:** King Ryu; **Exec.Prod:** Alan Fine, Joe Quesada, Dan Buckley, Cort Lane; **Marvel Animation/Atomic Cartoons**; 3.5

**Notes:** A series of shorts episodes which generally follow the format of Spider-Man interacting with kids, and then retelling an adventure featuring other heroes to ultimately teach the kids a life lesson.

**Ratings:** *IMDB: 57%*

**Review:** Clearly not intended for anyone above the age of 10, this is a simple series that does what it's supposed to do – teach kids important lessons using Marvel characters. It's entertaining and the kids will love it, and that's all it's intended to do.

## BIG HERO 6: THE SERIES
### (November, 2017 – February, 2021/Disney XD)

With the team assembled, Big Hero 6 find San Fransokyo coming under attack from a variety of villains, all scientifically enhanced. Although still studying, the team work to protect the city from these new threats.

**Regular Cast:** *Ryan Potter (Hiro Hamada), Scott Adsit (Baymax), Jamie Chung (Go Go Tomago), Genesis Rodriguez (Honey Lemon), Khary Payton (Wasabi), Brooks Wheelan (Fred),* Maya Rudolph (Aunt Cass), David Shaughnessy (Heathcliff), Alan Tudyk (Alistair Krei)

**Episodes:**

*1.1 Baymax Returns (Part 1)*
*1.2 Baymax Returns (Part 2)*
*1.3 Issue 188*
*1.4 Big Roommates 2*
*1.5 Fred's Bro-Tillion*
*1.6 Food Fight*
*1.7 Muirahara Woods*
*1.8 Failure Mode*
*1.9 Aunt Cass Goes Out*
*1.10 The Impatient Patient*
*1.11 Mr Sparkle Loses His Sparkle*
*1.12 Killer App*
*1.13 Small Hiro One*
*1.14 Kentucky Kaiju*
*1.15 Rivalry Weak*
*1.16 Fan Friction*
*1.17 Mini-Max*
*1.18 Big Hero 7*
*1.19 Big Problem*
*1.20 Steamer's Revenge*
*1.21 The Bot Fighter*
*1.22 Obake Yashiki*

*1.23 Countdown to Catastrophe (Part 1)*
*1.24 Countdown to Catastrophe (Part 2)*
*1.25 Countdown to Catastrophe (Part 3)*
*2.1 Internabout*
*2.2 Seventh Wheel*
*2.3 Prey Date*
*2.4 Something's Fishy*
*2.5 Nega-Globby*
*2.6 The Fate of the Roommates*
*2.7 Muira-Horror!*
*2.8 Something Fluffy*
*2.9 Supersonic Sue*
*2.10 Lie Detector*
*2.11 Write Turn Here*
*2.12 City of Monsters (Part 1)*
*2.13 City of Monsters (Part 2)*
*2.14 Mini-Maximum Trouble*
*2.15 El Fuego*
*2.16 The Globby Within*

**Notable Guest Cast:** Stan Lee (Fred's Dad), Dee Bradley Baker (Mayoi), Andy Richter (Globby), Mara Wilson (Liv Amara), Fred Tatasciore (Knox), James Cromwell (Callaghan), Gordon Ramsay (Bolton Gramercy), Jane Lynch (Supersonic Sue)

**Line Prod:** Jan Hirota; **Music:** Adam Berry; **Exec.Prod:** Nicholas Filippi, Mark McCorkle, Robert Schooley; **Ed:** Joseph Molinari, Charles T Jones; **Marvel Animation**; 23

**Notes:** The series follows on from the movie, though there are some changes made, which include ensuring that the team still have secret identities, and changing how they came up with their name. The most notable change, of course, is that the series is a more traditional 2D animation style, unlike the 3D style of the film. TJ Miller and Damon Wayans chose not to return to their roles, requiring Wheelan and Payton to replace them. For the first two seasons the episodes were 22 minutes long, but for the last season, the episodes were 11 minutes long, but then added to another episode to create the 22 minute run time. A number of short segments were also created in different styles. These were: **BAYMAX AND...** (May/June, 2018), **BAYMAX DREAMS** (September, 2018), **BIG CHIBI 6: THE SHORTS** (November, 2018 – April, 2019), **BAYMAX & MOCHI** (May, 2019), **DISNEY RANDOM RINGS** (June, 2020). The series was cancelled in January, 2021.

**Ratings:** *IMDB:* 72%

# MARVEL RISING
## (August, 2018 – December, 2019/Disney XD)

When evil starts to rear its head, Squirrel Girl and Ms Marvel join forces with Ghost-Spider, America Chavez and SHIELD Agents Quake and Patriot to stop it.

**Regular Cast:** *Chloe Bennet (Daisy Johnson/Quake), Kathleen Khavari (Kamala Khan/Ms Marvel), Kamil McFadden (Rayshun Lucas/Patiot), Milana Vayntrub (Doreen Green/Squirrel Girl), Tyler Posey (Dante Pertuz/Inferno), Cierra Ramirez (America Chavez), Dove Cameron (Gwen Stacy/Ghost-Spider), Dee Bradley Baker (Tippy-Toe), Sofia Wylie (Riri Williams/Ironheart)*

**Episodes:**

*1 Secret Warriors*
*2 Initiation*
*3 Chasing Ghosts*
*4 Heart of Iron*
*5 Battle of the Bands*
*6 Operation Shuri*
*7 Playing with Fire*

**Notable Guest Cast:** *Kim Raver (Captain Marvel), Roger Craig Smith (Captain America), Booboo Stewart (Exile), Ming-Na Wen (Hala the Accuser), Skai Jackson (Glory Grant), Steven Weber (Captain George Stacy), Tara Strong (Mary Jane Watson)*

**Prod:** Kalia Cheng Ramirez, Kenneth T Ito, Howard Schwartz; **Music:** Kristopher Carter, Michael McCuistion, Lolita Ritmanis; **Exec.Prod:** Greg Harman, Cort Lane, Eric Radomski, Joe Quesada, Dan Buckley; **Marvel Animation**; 22 - 80

**Notes:** *Marvel Rising* was pitched in 2017 as the next animation series, though the eventual decision was to release a series of animated movies (preceded by a set of shorts). The series reuses a lot of Marvel associated actors, but interestingly, Chloe Bennet reprises the role of Quake from *Agents of SHIELD*, and Milana Vayntrub played Squirrel Girl in the unscreened pilot of *New Warriors*. The song "*Born Ready*" was performed by Dove Cameron. *Initiation* was a collection of six four-minute short episodes, while *Secret Warriors* was a "movie" event. From there, the next five movies were announced over the following two years, though since the last one there has been no discussion on the series continuing.

**Ratings:** *IMDB:* 49%

**Review:** The real problem with *Marvel Rising* is that in its opening entries, it doesn't seem to know what it wants to be. With an animation style similar to *Spider-Man*, it has moments where it wants to be frivolous and fun like *Ultimate Spider-Man*, but then gets cold feet and becomes po-faced again, giving a strange bipolar feel to the series. However, by the time Gwen Stacy joins the team, the characters have a more teenage interaction, and the series seems to know its target audience, delivering some enjoyable tales to introduce interesting characters, even if the stories themselves aren't too deep.

## *M.O.D.O.K.*
### *(May, 2021/Hulu)*

Between attempting to get AIM back from a multi-national buy out and stopping a time-travelling version of himself, MODOK finds himself struggling to keep his family together after accidentally taking over his wife's body.

**Regular Cast:** *Patton Oswalt (MODOK),* Melissa Fumero (Melissa Tarleton), Aimee Garcia (Jodie Ramirez-Tarleton), *Wendi McLendon-Covey (Monica Rappaccini/Scientist Supreme),* Ben Schwartz (Louis Tarleton)

**Episodes:**

*1 If This Be...MODOK!*
*2 The MODOK That Time Forgot!*
*3 Beware What From Portal Comes!*
*4 If Saturday Be...For the Boys!*
*5 If Bureaucracy Be Thy Death!*
*6 Tales from the Great Bar-Mitzvah War!*

*7 This Man...This Makeover!*
*8. O, Were Blood Thicker Than Robot Juice!*
*9. What Menace Doth the Mailman Deliver!*
*10. Days of Future MODOKs*

**Notable Guest Cast:** Beck Bennett (Austin Van Der Sleet), *Jon Daly (Super-Adaptoid),* Sam Richardson (Gary), *Jon Hamm (Iron Man), Nathan Fillion (Wonder Man), Whoopi Goldberg (Poundcakes), Alan Tudyk (Arcade), Bill Hader (The Leader/Angar the Screamer),* Kevin Michael Richardson (Mr Sinister/Mandrill), *Meredith Salenger (Madame Masque), Chris Parnell (Tenpin), Eddie Pepitone (Melter), Jonathan Kite (Tatterdemalion)*

**Dir:** Eric Towner, Alex Kamer, Paula Haifley; **Ed:** Chris Rogers; **Prod:** Itai Grunfeld, Geoff Barbanell; **Music:** Daniel Rojas; **Exec.Prod:** Brett Cawley, Robert Maitia, Grant Gish, Joe Quesada, Karim Zreik, Jeph Loeb, Patton Oswalt, Jordan Blum; **Created & Written by** Jordan Blum & Patton Oswalt; **DOP:** Susie Shircliff; **Multiverse Cowboy/Stoopid Buddy Stoodios/Marvel Television**; 25

**Notes:** *MODOK* was to be one of four different series (the others being *Howard The Duck* – written by Kevin Smith and Dave Willis; *Hit-Monkey* – see below; and *Tigra & Dazzler* – written by Erica Rivinoja and Chelsea Handler) that would culminate in *The Offenders*. They were announced in February, 2019, but in January, 2020 all but *MODOK* and *Hit-Monkey* were abandoned. When Marvel Television was merged into Marvel Studios, Kevin Feige was happy for the series to continue. The decision was taken to film the series using stop-motion animation. It was confirmed in May, 2022 there would not be a second series.

**Ratings:** *IMDB:* 63%

**Review:** *MODOK* is an entertaining enough premise, though rather than play into the idea that Modok is an insane villain, he's presented as pretty useless. And while arguably he achieves his goal, the idea of a villain realising he really loves his family has sort of already been done in **DESPICABLE ME**. There are laugh out loud moments, and a lot of easter eggs for fans, but the casual viewer probably won't get much out of this one.

## *SPIDEY AND HIS AMAZING FRIENDS*
### *(August, 2021 –/Disney Junior)*

Spider-Man and his friends, Ghost-Spider and Spin, take on the villains of New York, joining forces with other superheroes to keep the people safe.

**Regular Cast:** *Banjamin Valic (Peter Parker/Spider-Man), Jakari Fraser (Miles Morales/Spin), Lily Sanfelippo (Gwen Stacy/Ghost-Spider),* Nicholas Roye (WEB-STER), Dee Bradley Baker (TRACE-E/TWIST-E/TWIRL-E), *Melanie Minichino (Aunt May), Gabrielle Ruiz (Rio Morales), Eugene Byrd (Jeff Morales), Kari Wahlgren (Helen Stacy), Scott Porter (George Stacy)*

**Episodes:**

**Notable Guest Cast:** *Sandra Saad (Captain Marvel), Tru Valentino (Black Panther), Armen Taylor (Hulk), John Stamos (Tony Stark/Iron Man), Sean Giambrone (Ant-Man), Maya Tuttle (Wasp), Kelly Ohnanian (Doc Ock), JP Karliak (Green Goblin), Justin Shenkarow (Rhino), Stephanie Lemelin (Electro), Jaiden Klein (Black Cat), Thomas F Wilson (Sandman),* Bindi Irwin (Isla Coralton)

**Dir:** Darren Bachynski; **Prod:** Steven L Grover, Ashely Rideout; **Music:** Patrick Stump; **Exec.Prod:** Harrison Wilcox; **Ed:** Tristan Goodes; **Marvel Entertainment/Atomic Cartoons/Marvel Studios**; 22

**Notes:** A more child friendly television series that uses CG animation in a more in the style of *PAW PATROL,* and aimed at the audience. As such, the age of the spider heroes has been scaled down, although most of the other superheroes are still adults. Strangely the villains ages have been scaled down as well. The trio have a headquarters in the Parker's backyard (though the Team Transport vehicle rises up out of the ground, destroying the secret, somewhat), and all three have mini spider-computers that assist them, as well as their own personal vehicles. The second season gains the subtitle *Glow Webs Glow.*

**Ratings:** *IMDB:* 58%

**Review:** It's not really fair to review this, as the audience for this series is definitely not the typical fan. Bright and colourful, often with an important message for kids to learn, the series is lots of fun and makes great use of the Marvel characters available to it. Children will no doubt love it. Anyone over the age of 10, however, will probably have little tolerance for it. But as they aren't the intended audience, it doesn't matter. For what it is, the show succeeds.

## *HIT MONKEY*
### *(November, 2021 /Hulu)*

When jaded hitman Bryce Fowler is killed in front of a monkey pack, just before the pack itself is shot to death, the only surviving monkey finds himself bonded to Fowler and sets out to take revenge on both their enemies, using Fowler's skills and weapons. Meanwhile, the forthcoming Japanese election is tied into the very group that Hit Monkey is hunting.

**Regular Cast:** Ally Maki (Haruka), *Olivia Munn (Akiko Yokohama),* Nobi Nakanishi (Ito), *Fred Tatasciore (Hit Monkey),* George Takei (Shinji Yokohama) and Jason Sudeikis (Bryce Fowler)

**Episodes:**

1. *Pilot*
2. *Bright Lights, Big City*
3. *Legend of the Drunken Monkey*
4. *The Code*
5. *Run Monkey, Run*

6. *The Long Goodbye*
7. *Sayonara Monkey!*
8. *Home, Sweet Home*
9. *The End: Part One*
10. *The End: Part Two*

**Notable Guest Cast:** *Reiko Aylesworth (Lady Bullseye), Noshir Dalal (Silver Samurai)*

**Dir:** Neal Holman; **Writers:** Josh Gordon, Will Speck, Keith Foglesong, Duffy Boudreau, Matteo Borghese, Rob Turbovsky, Albertina Rizzo, Ken Kobayashi, Paul Levitt; **Ed:** Nate Ladd, Brad Lee Zimmerman; **Prod:** Duffy Boudreau, Molly Brock, Marcus Rosentater; **Music:** Daniel Rojas; **Exec.Prod:** Josh Gordon, Will Speck, Grant Gish, Jeph Loeb; **Created by** Josh Gordon; **Speck Gordon Inc/Floyd County Productions/Marvel Television**; c 25

**Notes:** *Hit-Monkey* was another of the series that was to potentially lead to *The Offenders*. Announced in February, 2019 the series was far enough into production that when Marvel Studios absorbed Marvel Television, the project continued. The series was renewed for a second season in 2023.

**Ratings:** *IMDB:* 78%

**Review:** *HIT MONKEY* takes a little time to find its feet, and though the cast is good, the premise sometimes struggles to deliver. Despite this, Sudeikis' performance seems oddly pitched, which isn't helped by the fact he only ever interacts with Hit Monkey. The conclusion is satisfying and the story ultimately works, but the hints in the last episode for the possible second season make it seem far more interesting than this first season.

## *BAYMAX!*
### *(June, 2022/Disney+ XD)*

Several people encounter painful incidents, but are helped by Baymax, though not usually to great effect.

**Regular Cast:** *Scott Adsit (Baymax), Ryan Potter (Hiro Hamada)*

**Episodes:**

| | |
|---|---|
| *1.1 Cass* | *1.4 Mbita* |
| *1.2 Kiko* | *1.5 Yachi (Part 1)* |
| *1.3 Sofia* | *1.6 Baymax (Part 2)* |

**Notable Guest Cast:** Maya Rudolph (Aunt Cass), Dichen Lachman, Sarah-Nicole Robles

**Dir:** Dean Wellins, Dan Abraham, Mark Kennedy, Lissa Treiman; **Writers:** Cirocco Dunlap, Julia Fitzmaurice, Jeff Snow; **Music:** Dominic Lewis; **Exec.Prod:** Roy Conli, Don Hall, Jennifer Lee; **Ed:** Sarah K Reimers, Shannon Stein; **Walt Disney Animation Studios**; 10

**Notes:** A spin-off from *BIG HERO 6*, though curiously made by Walt Disney Animation Studios as Marvel Animation had been folded into Marvel Studios. Notably the show has impressive diversity, though this has come with the expected controversy as well.

**Ratings:** *IMDB:* 69%

## *MOON GIRL AND DEVIL DINOSAUR*
### *(February, 2023/Hulu)*

Problems in her neighbourhood convince child genius Lunella Lafayette to go out and do something about it. Her new friend Casey suggests she adopts the identity of Moon Girl, and with the assistance of a dinosaur that Lunella accidentally manages to bring to the present, the pair go out and defend their neighbourhood from the supervillains that infest it.

**Regular Cast:** *Diamond White (Lunella Lafayette/Moon Girl), Fred Tatasciore (Devil Dinosaur, Devos the Devastator),* Alfre Woodard (Miriam "Mimi" Lafayette), Sasheer Zamata (Adria Lafayette, Flying Fox), Jermaine Fowler (James Lafayette Jr), Gary Anthony Williams (James "Pops" Lafayette, Rockin' Rudy, Garko the Man-Frog), Libe Barer (Casey Calderon), *Laurence Fishburne (The Beyonder, Bill Foster)*

**Episodes:**

1. *Moon Girl Landing*
2. *The Borough Bully*
3. *Run the Rink*
4. *Check Yourself*
5. *Hair Today, Gone Tomorrow*
6. *The Beyonder*
7. *Moon Girl's Day Off*
8. *Teacher's Pet*
9. *Skip This*
10. *Goodnight, Moon Girl*
11. *Like Mother, Like Moon Girl*
12. *Today, I Am Woman*
13. *Devil on her Shoulder*
14. *Coney Island Baby!*

**15. OMG Issue #1**

**16. OMG Issue #2**

**Notable Guest Cast:** Omid Abtahi (Ahmed), Utkarsh Ambudkar (Anand), Michael Cimino (Eduardo), Indya Moore (Brooklyn), Craig Robinson (Prinicpal Nelson), Ian Alexander (Tai), Tajinae Turner (Geri), *Allison Brie (Aftershock),* Jennifer Hudson (Mane), *Maya Hawke (Abyss), Anna Akana (Odessa Drake), Cobie Smulders (Maria Hill),* Wesley Snipes (Maris Morlak), *Method Man (Torg),* Luis Guzmán (Principal Peña), May Calamawy (Fawzia), Daveed Diggs (Rat King), Kari Wahlgren (Linh Pham)

**Dir:** Trey Buongiorno, Christine Liu, Ben Juwono, Rodney Clouden, Samantha Suyi Lee; **Writers:** Jeffrey M Howard, Kate Kondell, Halima Lucas, Maggie Rose, Lisa Muse Bryant, Liza Hara, Taylor Vaughn Lasey; **Ed:** Sandra Powers; **Prod:** Pilar Flynn; **Music:** Raphael Saadiq; **Exec.Prod:** Steve Loter, Laurence Fishburne, Helen Sugland; **Cinema Gypsy Productions/Disney Television Animation/Marvel Animation**; 22

**Notes:** Raphael Saadiq provides a new song for every episode. The title song – Moon Girl Magic – was also written by him (with Halima Lucas and Taura Stinson) and performed by Diamond White. Laurence Fishburne, who was the primary impetus for the project, wanted the series to look like a moving comic book, and the animation style draws inspiration from both the Spiderverse movies and Marvel's **WHAT IF…?**. Much of the series was made remotely due to the COVID-19 pandemic. The series was renewed for a second season in October, 2022, before the first season has been aired.

**Ratings:** *IMDB:* 78%

**Review:** There is a bundle of energy in this series, kicking off with a title sequence that is both colourful and has a brilliant theme song. Off to such a great start, it's difficult to go wrong, and the series doesn't fail to impress with some very effective casting and entertaining storylines. It's not mired down in moodiness, but rather tries to be a show that children will invest in, and on that level it succeeds brilliantly. The fact that it showcases a very different kind of hero is something that works very much to its strengths.

# Coming Soon…

## X-MEN '97
### (2024/Disney+)

Now under the leadership of Magneto, the X-Men continue to build human/ mutant relationships by saving a lot of people from much evil.

**Regular Cast:** *Ray Chase (Scott Summers/Cyclops), Chris Potter (Gambit), Jennifer Hale (Jean Grey), Adrian Hough (Nightcrawler), Lenore Zann (Rogue), Cal Dodd (Woverine), Alison Sealy-Smith (Storm), George Buza (Henry McCoy/Beast), Holly Chou (Jubilee), Christopher Britton (Mr Sinister), Matthew Waterson (Magneto), Gui Agustini (Roberto DaCosta/Sunspot)*

**Notable Guest Cast:** Alyson Court, Catherine Disher, Chris Potter, Adrian Hough, Anniwaa Buachie, JP Karliak, Jeff Bennett

**Exec.Prod:** Kevin Feige, Dana Vasquez-Eberhardt, Brad Winderbaum, Beau DeMayo; **Marvel Studios Animation**

**Notes:** A continuation of the 1990's *X-Men* animated series. Court, Disher, Potter and Hough all return from the previous version, though voicing new characters. The series was delayed due to the Hollywood strikes in 2023.


## SPIDER-MAN: FRESHMAN YEAR
### (2024/Disney+)

Having recently become Spider-Man, Peter Parker attempts to fight crime and balance his school life.

**Regular Cast:** *TBC*

**Notable Guest Cast:** *Charlie Cox (Daredevil)*

**Exec.Prod:** Kevin Feige, Brad Winderbaum, Jeff Trammell; **Walt Disney Animation Studios/Marvel Studios Animation**

**Notes:** The newest Spider-Man animated series was originally pitched as a series that would focus on Peter before the events of **CAPTAIN AMERICA: CIVIL WAR**, but by the time discussion of this took place at SDCC '22, it became clear that ties to the MCU had been abandoned, not least because Peter would be interacting with Dr Strange, Dr Octopus and the Green Goblin well before he meets any of them in the MCU. Despite this, it was confirmed that Charlie Cox would be returning to voice Daredevil. Amadeus Cho, Harry Osborn, Nico Minoru and May Parker would be regular characters, while other villains include Scorpion, Chameleon, Rhino, Tarantula and Unicorn.

# V. LIVE ACTION TELEVISION SERIES

## SPIDEY SUPER STORIES
### (1974 – 1977/PBS)

Spider-Man stops crooks on a weekly basis from carrying out their dastardly plans.

**Regular Cast:** *Danny Seagren (Spider-Man)*
**Notable Guest Cast:** Morgan Freeman (Count Dracula)

*The Electric Company* **Prod:** Naomi Foner, Andrew B Ferguson Jr, Edith Zornow; **Music:** Gary William Friedman; **Exec.Prod:** David D Connell, Joan Ganz Cooney, Samuel Y Gibbon Jr, Walt Rauffer; **Prod.Des.:** Ron Baldwin, Bill Bohnert; **Costumes:** Ramsay Mostoller; **Children's Television Workshop**; 5

**Notes:** The series was actually a five minute skit that featured during every episode of the television show *The Electric Company*, and outside of Spider-Man, didn't really make much use of the Spider-Man mythology – indeed, Peter Parker never appears for instance. Aimed more at a child audience, it did use comic book themes – Spider-Man spoke using word bubbles for instance.

**Ratings:** *IMDB:* 83%

## THE INCREDIBLE HULK
### (November, 1977 – June, 1982/CBS)

David Banner, haunted by the death of his wife, works on a theory that people can access hidden strength in extreme emotional situations, stimulated by gamma radiation. He tests the theory on himself, but the gamma dose is far too high, and as a result, his emotion can trigger a transformation into the Hulk. On the run, Banner drifts from town to town giving help where possible, always aware the authorities, and journalist Jack McGee, are one step behind him.

**Regular Cast:** *Bill Bixby (Dr David Banner)*, Jack Colvin (Jack McGee), *Lou Ferrigno (The Incredible Hulk)*

**Episodes:**

| | |
|---|---|
| *1.1 The Incredible Hulk (Pilot)* | *1.4 The Beast Within* |
| *1.2 Death in the Family* | *1.5 Of Guilt, Models and Murder* |
| *1.3 Final Found* | *1.6 Terror in Times Square* |

**Prod:** Kenneth Johnson, Nicholas Corea, Robert Bennett Steinhauer, Jill Donner, Karen Harris, James G Hirsch, James D Parriott, Jeff Freilich, Andrew Schneider; **Music:** Joseph Harnell, C R Cassey; **Exec.Prod:** Kenneth Johnson; **DOP:** John McPherson, Eddie Ro Rotunno, Vincent A Martinelli, Howard Schwartz, Charles W Short; **Costumes:** Charles Waldo [1.1-2.2], George Whittaker [2.3-2.7], Brienne Glyttov [2.8-5.7]; **Universal Television**; c 50

**Notes:** Jack Kirby, co-creator of the Hulk, makes a cameo appearance in 2.18. The first two episodes were feature length pilots of the series, but outside of America, were actually released in cinemas, though "Death in the Family" was named "The Return of the Hulk". Ted Cassidy provides the opening narration, as well as the voice of the Hulk, uncredited. Richard Kiel was originally cast as the Hulk, but was considered not bulky enough, and so Ferrigno was cast to take over. Arnold Schwarzenegger was also considered. The series was cancelled, more as a cost cutting exercise, rather than any belief the ratings were suffering. Bruce Banner's name was changed to David because Kenneth Johnson hated alliterative names. At one point there was a plan to introduce, and spin-off, a female Hulk; Banner's sister. Marvel got wind of this and immediately created She-Hulk to get the rights to such a character.

**Ratings:** *IMDB:* 70%

## THE AMAZING SPIDER-MAN
### *(April, 1978 – July, 1979/CBS)*

Newly empowered Peter Parker decides to fight crime in New York City in the guise of Spider-Man.

**Regular Cast:** *Nicholas Hammond (Peter Parker/Spider-Man), Robert F Simon (J Jonah Jameson),* Chip Fields (Rita Conway), Michael Pataki (Capt Barbera)[1.1-1.5], Ellen Bry (Julie Masters)[2.1-2.8]

**Episodes:**

| | |
|---|---|
| *1.1 Deadly Dust, Part 1* | *2.2 A Matter of State* |
| *1.2 Deadly Dust, Part 2* | *2.3 The Con Caper* |
| *1.3 The Curse of Rava* | *2.4 The Kirkwood Haunting* |
| *1.4 Night of the Clones* | *2.5 Photo Finish* |
| *1.5 Escort to Danger* | *2.6 Wolfpack* |
| *2.1 The Captive Tower* | *2.7 The Chinese Web, Part 1* |

### 2.8 The Chinese Web, Part 2

**Notable Guest Cast:** Rosalind Chao (Emily Chan), Theodore Bikel (Mandak), Ted Danson (Major Collings), Morgan Fairchild (Lisa Benson), *Irene Tedrow (Aunt May Parker),* Madeleine Stowe (Maria Calderon)

**Prod:** Robert James, Ron Satlof, Lionel E Siegel; **Music:** Dana Kaproff, Stu Phillips; **Exec.Prod:** Charles W Fries, Daniel R Goodman; **DOP:** Jack Whitman, Vincent A Martinelli; **Costumes:** Frank Novak; **Charles Fries Productions/Dan Goodman Productions/Danchuk Productions**; 45

**Notes:** The television series launched from the 1977 television movie, wasn't entirely supported by CBS, who, rather than committing to a series proper, instead commissioned groups of episodes which they scheduled at various times throughout the year. The second season was toned to aim at a more adult market, but the ratings for that demographic never really improved, and with the cost and the fact CBS was becoming known as a "superhero network", the decision was taken to cancel it. There were plans, at one point, to include the cast in one of the 1980's *Incredible Hulk* telemovies, but the plans were dropped.

**Ratings:** *IMDB:* 65%

## SPIDER-MAN (スパイダーマン)
### *(May, 1978 – March, 1979/Tokyo Channel 12)*

Takuya Yamashiro comes across "The Marveller", a crashed spaceship from the planet Spider. Joined by his father, space archaeologist Hiroshi, they learn that Garia is the sole survivor of his race, and he was hunting down Professor Monster and his Iron Cross Army. Hiroshi is killed, and Garia dies, but not before injecting Takuya with his blood, giving him spider-like abilities, as well as access to the Marveller – which can become the giant robot Leopardon. Adopting the name Spider-Man, Takuya sets out to stop Professor Monster.

**Regular Cast:** Shinji Tōdō (Takuya Yamashiro), Mitsuo Andā (Professor Monster), Yukie Kagawa (Amazoness), *Hirofumi Koga (Spider-Man),* Tōru Ōhira (Narrator), Izumi Oyama (Shinko Yamashiro), Rika Miura (Hitomi Sakuma), Yoshiharu Yabuki (Takuji Yamashiro)

**Episodes:**

*1.1 The Time of Revenge Has Come! Beat Down Iron Cross Group!!*

*1.2 Mysterious World! The Man Who Follows His Fate*

*1.3 Mysterious Thief 001 VS. Spider-Man*

*1.4 The Terrifying Half Merman! The Miracle-Calling Silver Thread*

*1.5 Crash Machine GP-7! The Oath Siblings*

*1.6 Shuddering Laboratory! Devilish Professor Monster*

*1.7 Fearful Hit Tune! Song Dancing Murder Rock*

*1.8 A Very Mysterious Folktale: The Cursed Cat Mound*

*1.9 Motion Accessory is a Loveful Beetle Insect Spy*

*1.10 To the Flaming Hell: See the Tears of the Snake Woman*

*1.11 Professor Monster's Ultra Poisoning*

*1.12 Becoming Splendid: To the Murderous Machine of Transformation*

*1.13 The Skull Group VS. The Devilish Hearse*

*1.14 Giving Father! Fight to the Song of the Hero*

*1.15 The Life of Our Arrangement*

*1.16 Fine Dog! Run to the Under of Father*

*1.17 Pro Wrestler Samson's Tears*

*1.18 In the Mother's Chest: Resurrect the Young Boys*

*1.19 The Boy Phantom: To the Villageless Map*

*1.20 Riddle: Calling the Riddle of My Secret Birth*

*1.21 Fall to the Great Skies: Father's Love*

*1.22 Shedding Tears to the Dark Fate: Father and Child*

*1.23 To the Love Academy of the Homeless Children*

*1.24 Cockroach Boy: Great War*

*1.25 Treasure, Dog, and Double Grow Human*

*1.26 To the Absolute Crisis: The Imitation Hero*

*1.27 Farewell War Buddy: Beloved German Shepherd*

*1.28 The Front of the Alley: Boys' Detective Group*

*1.29 Hurry, GP-7: Time of Stop Sign*

*1.30 Good Luck, Beautiful Police Officer*

*1.31 There Is No Child-Taking Detective Tomorrow*

*1.32 Sweet Whispering Enchantress*

*1.33 The Boy Teases the Horrible Wild Girl*

*1.34 Surprising Camera: Murderous Event*

*1.35 From the Unexplored Amazon: Here Comes the Mummified Beautiful Woman*

*1.36 The Onion Silver Mask and the Boys' Detective Group*

*1.37 From the Secret Messenger of Hell: Great King Enma*
*1.38 The First Tin Plate Evening Star and the Boys' Detective Group*
*1.39 Sports World: One Great Meeting*
*1.40 Farewell Zero Battle Tricks*
*1.41 The Hero's Shining Hot Blood*

**Prod:** Susumu Yoshikawa, Hiroshi Ishikawa; **Music:** Michiaki Watanabe; **Toei/ Toei Advertising/Tokyo Channel 12**; 24

**Notes:** Outside of the Spider-Man concept, there is literally nothing connecting the series to the comics, though this was not the plan producers Yoshikawa and Ishikawa had. Originally this was going to be faithful to the comic, but they were forced to include a giant robot. The success of the series led to the revival of the Super Sentai Series (costumed heroes riding giant robots) and a third series was planned for Captain America. He would be Captain Japan, however, though as the plans moved forward, he became Battle America, and was joined in his robot by Battle France, Battle Kenya and Battle Cossack. Oddly enough, Toei's deal with Marvel actually granted them access to any Marvel characters they wanted. The comic *Spider-Verse* saw many alternate Spider-Men come together, and one was indeed Takuya Yamashiro with his Leopardon.

**Ratings:** *IMDB:* 67%

## POWER PACK
### (1991)

The Power family adapt to a new life after a move, but Jack soon becomes involved with a dare to visit the mansion of Dr Mobius. Jack steals a medallion, unwittingly releasing the ghost of Mobius who wants his medallion back.

**Cast:** *Nathaniel Moreau (Alex Power), Margot Finley (Julie Power), Bradley Machry (Jack Power), Jacelyn Holmes (Katie Power), Jonathan Whittaker (Dr James Power), Cheryl Wilson (Margaret Power),* Daniel DeSanto (Eddie), Christian Masten (Harlan), Rachel Wilson (Tina), Charlene DiPardo (Rhonda), Greg Swanson (Mobius), Rudy Webb (Teacher), Lana Goldstein (Girl)

**Dir:** Rick Bennett; **Writer:** Jason Brett; **Prod:** Richard Borchiver; **Exec.Prod:** Jon Slan; **DOP:** Bert Dunk; **Prod.Des:** Michael Borthwick; **Ed:** Dave Goard; **Costumes:** Annie Nikolajevich; **New World Entertainment/Marvel**

**Enterprises**; 27

**Notes:** The first New World attempt at a Marvel television series, though only a pilot was made as the series was never picked up.

**Comic Notes:** The Power family first appeared in *Power Pack #1 (May, '84)*, and were created by Louise Simonson and June Brigman. Generally the pack's superpowers have remained as per the comic book, though in the comic Julie's powers are more to do with acceleration and she can fly. Jack's powers are completely different – in the comics he can control his density, not his size. One other notable change from the comic is that James and Margaret are aware of their children's abilities, something which doesn't happen in the comic.

**Ratings:** *IMDB:* 57%

**Review:** It's not difficult to see why this wasn't picked up for series, as for the most part it is awful. The actors playing the children are not remotely appealing, and the script is lazy, even for a children's program. The Power Pack are a great concept, but this show doesn't come close to capturing the wit or charm of the comic.

## *MUTANT X*
### *(October, 2001 – May, 2004)*

A company called Genomex once experimented on human embryos, and years later, the children who have resulted from those experiments display surprising powers. While Mason Eckhart – head of security at Genomex – seeks to hunt these mutants down to use them, former chief biogeneticist, Adam Kane, starts an underground railway to protect the new mutants.

**Regular Cast:** Karen Cliche (Lexa Pierce)[3.1-3.22], *Forbes March (Jesse Kilmartin), Victoria Pratt (Shalimar Fox), Lauren Lee Smith (Emma DeLauro)* [1.1-2.22], *Victor Webster (Brennan Mulwray) and John Shea as Adam*[1.1-2.22*]
*Guest stars in season three.*

**Episodes:**

*1.1 The Shock of the New*
*1.2 I Scream the Body Electric*
*1.3 Russian Roulette*

*1.4 Fool for Love*
*1.5 Kilohertz*

**Notable Guest Cast:** Tom McCamus (Mason Eckhart), Andrew Gillies (Dr Kenneth Harrison), Michael Easton (Gabriel Ashlocke), and George Buza (Lexa's Dominion Contact)

**Prod:** John Ledford, Jonathan Hackett, Jamie Paul Rock; **Music:** Louis Natale [1.1-1.22], Donald Quan [2.1-3.22]; **Exec.Prod:** John Ledford, Peter Mohan, Avi Arad, Jay Firestone, Adam Haight, Rick Ungar, Seth Howard; **DOP:** Colin

Hoult, Nikos Evdemon, Alwyn Kumst, Jim Westenbrink; **Prod.Des.:** John Blackie, Rocco Matteo; **Costumes:** Terri de Haan, Steven Wright, Laurie Drew; **Fireworks Entertainment/Tribune Entertainment/Marvel Entertainment/Mutant X Productions Ltd**; 45

**Notes:** This was essentially Marvel's attempt to make an X-Men television series without having access to the X-men, though ultimately that would lead to a lot of legal troubles. The series was essentially made by Fireworks Entertainment, and the decision to end it came from the company's sale to ContentFilm. It was surprisingly popular, and Marvel even went so far as to do some comic work based on the series. However, Fox immediately realised what was going on and filed suit against Marvel, Fireworks and Tribune. Marvel countersued to keep production going, and won, but it was made clear X-Men were not to be referenced in any way in the series. In 2003, Fox and Marvel settled out of court, though Fox still kept its suit against Fireworks and Tribune. Tribune consequently sued Marvel for encouraging them to use X-Men material that required enormous man-hours to remove. It's fair to say that no one really came out a winner from this one.

**Ratings:** *IMDB:* 60%

**Review:** I have a lot of time for Victoria Pratt and John Shea, so I did warm to this series thanks to them, but it's difficult to get rid of the nagging feeling that this is just a cut-price X-Men. The concept of the underground railway is a good one but it never quite sparks in the way that it should. This is not a series that is mind blowing, but it is a good way of passing time.

## BLADE: THE SERIES
### *(June – September, 2006/Spike)*

Krista Starr returns from Iraq, only to learn that her brother was a vampire familiar who was killed by the vampire Marcus Van Sciver. Van Sciver hopes to find an antidote to vampire weaknesses, but falls for Krista and tries to turn her. Krista, however, meets Blade, who she agrees to work with in order to bring Van Sciver down.

**Regular Cast:** *Kirk "Sticky" Jones (Blade),* Jill Wagner (Krista Starr), Nelson Lee (Shen), Jessica Gower (Chase) and Neil Jackson as Marcus Van Sciver

**Episodes:**

**Prod:** Gordon Mark; **Music:** Ramin Djawadi; **Exec.Prod:** Jim Rosenthal, Jon Kroll, Ari Arad, Avi Arad, David Simkins [1.3-1.13], David S Goyer; **DOP:** Robert New; **Prod.Des.:** Eric Fraser; **Costumes:** Kate Main, Katia Stano; **Spike TV/ Marvel Enterprises/New Line Television/Phantom Four**; 45

**Notes:** Originally to be produced by Showtime, star Wesley Snipes and be based on *Blade: The Vampire Hunter*, once Snipes declined, Showtime passed. Spike TV picked it up, but elected to just play the series as a loose sequel to the movies. Although there is mention of **Blade: Trinity**, there is no sign of Hannibal King or Abby Whistler. The pilot doesn't have "The Series" as a subtitle, but rather uses "House of Chthon". Also, a character in the pilot references Marc Spector – the Moon Knight.

**Ratings:** *IMDB:* 66%

**Review:** There's something that doesn't quite work with this series. Jones does his best as Blade, but he lacks the menace and unpredictability of Snipes. There's sexiness, violence and gore, but it all seems a bit lacklustre, and at times just boring. It tries very hard, but it doesn't ever hit the mark.

## *LEGION*

### *(February, 2017 – 2019/FX)*

Stuck in a mental institute for attempting to commit suicide, David Haller meets Syd Barrett and falls in love, though when they finally kiss for the first time, she swaps bodies with him, and he is able to leave the institute leading him to an organisation lead by Melanie Bird. David is the son of Charles Xavier who, many years ago confronted Amahl Farouk and defeated the powerful telepath by planting him in his son's mind. But the Shadow King is merely sitting in wait, and David will be forced to confront the demon and his own past to decide on his future.

**Cast:** *Dan Stevens (David Haller),* Rachel Keller (Syd Barrett), Aubrey Plaza

(Lenore "Lenny" Busker), Bill Irwin (Cary Loudermilk), *Navid Negahban (Amahl Farouk/The Shadow King)\**, Jermaine Clement (Oliver Bird)^^, Jeremie Harris (Ptonomy Wallace), Amber Midthunder (Kerry Loudermilk), Hamish Linklater (Clark Debussy)^ with Katie Aselton (Amy Haller) and Jean Smart (Melanie Bird)\*\* and Lauren Tsai (Jia-yi/Switch)\*\*\*

\*Regular for seasons two and three. \*Regular for seasons one and two, recurring season three. \*\*\* Regular season three only. ^Recurring season one, regular seasons two and three. ^^Recurring seasons one and three, regular season two.

**Episodes:**

| | |
|---|---|
| *1.1 Chapter 1* | *2.7 Chapter 15* |
| *1.2 Chapter 2* | *2.8 Chapter 16* |
| *1.3 Chapter 3* | *2.9 Chapter 17* |
| *1.4 Chapter 4* | *2.10 Chapter 18* |
| *1.5 Chapter 5* | *2.11 Chapter 19* |
| *1.6 Chapter 6* | *3.1 Chapter 20* |
| *1.7 Chapter 7* | *3.2 Chapter 21* |
| *1.8 Chapter 8* | *3.3 Chapter 22* |
| *2.1 Chapter 9* | *3.4 Chapter 23* |
| *2.2 Chapter 10* | *3.5 Chapter 24* |
| *2.3 Chapter 11* | *3.6 Chapter 25* |
| *2.4 Chapter 12* | *3.7 Chapter 26* |
| *2.5 Chapter 13* | *3.8 Chapter 27* |
| *2.6 Chapter 14* | |

**Notable Guest Cast:** *Harry Lloyd (Charles Xavier), Stephanie Corneliussen (Gabrielle Xavier), Saïd Taghmaoui (Amahl Farouk)*, David Selby (Brubaker), Ellie Araiza (Philly), Mackenzie Gray (The Eye), Scott Lawrence (Henry Poole), Brad Mann (Rudy), Devyn Dalton (The Angry Boy), Qinton Boisclair (The Devil with the Yellow Eyes), Marc Oka (Admiral Fukyama), Samantha Cormier (Cynthia), Nathan Hurd (The Monk), Jelly Howie, Brittney Rose Parker, Lexa Gluck, Rachele Schank, Marikah Cunningham, Tiffany Feese (The Vermillion), Jason Mantzoukas (The Wolf), Jon Hamm (Narrator)

**Prod:** Brian Leslie Parker [1.1-1.8], Regis Kimble, Erik Holmberg [2.1-3.8], Craig Yahata [2.1-3.8], Ben H Winters [3.1-3.8]; **Music:** Jeff Russo; **Exec.Prod:** Noah Hawley, John Cameron, Lauren Schuler Donner, Bryan Singer, Simon Kinberg, Steve Blackman, Jeph Loeb, Jim Chory, Alan Fine, Stan Lee, Joe Quesada, Karim Zreik; **Created by** Noah Hawley; **DOP:** Dana Gonzales [1.1-1.8], Polly Morgan [2.1-2.11], Alexi Disenhof [3.1-3.8], Erik Messerschmidt [3.1-3.8]; **Prod.Des.:** Michael Wylie [1.1-2.11], Marco Niro [3.1-3.8]; **Costumes:** Carol Case [1.1-1.8],

Robert Blackman [2.1-3.8]; **26 Keys Productions/Donners Company/Bad Hat Harry/Kinberg Genre/Marvel Television/FX Productions**; 45 - 70

**Notes:** One of two television shows ordered to pilot after Fox and Marvel reached a surprising agreement to co-produce the series together. Hawley, Peter Calloway, Jennifer Yale and Nathaniel Halpern wrote season one, while Hawley wrote all the episodes of season two with some co-written with Halpern or Jordan Crair. Hawley also directed. Despite the success of the series, Hawley only ever envisioned three seasons, so the Fox buyout had no impact on the series end. Syd Barrett is named for the Pink Floyd guitarist who also suffered from mental illness. Lauren Tsai made her acting debut in season three; she is also an illustrator with Marvel. Rather amusingly, instead of "Previously on..." for the recap montages, the voice overs use alternative phrases such as "Ostensibly on..." or "Apparently on..."

**Review:** Marvel television tends to be on fire with its MCU material, but divorced from that and partnering with Fox, it proves that at its heart, Marvel TV simply know how to make excellent television – *Legion* is quite simply brilliant. Brilliant performances all round, the show is dense and complex but utterly rewarding. Additionally, some of the creative decisions are unexpected, making for addictive viewing.

## *THE GIFTED*
### *(October, 2017 - February, 2019/Fox)*

Stuck in a mental institute for attempting to commit suicide, David Haller meets Syd Barrett and falls in love, though when they finally kiss for the first time, she swaps bodies with him, and he is able to leave the institute. Syd gains David's abilities, and kills everyone in the institute, including Haller's best friend Lenny. Soon, Syd catches up with David and he is brought before Melanie Bird, who has a plan for him, and his unique abilities.

**Cast:** Stephen Moyer (Reed Strucker), Amy Acker (Caitlin Strucker), Sean Teale (Marcos Diaz/Eclipse), Natalie Alyn Lind (Lauren Strucker), Percy Hynes White (Andy Strucker), Coby Bell (Jace Turner), *Jamie Chung (Clarice Fong/ Blink), Blair Redford (John Proudstar/Warpath), Emma Dumont (Lorna Dane/ Polaris), Skyler Samuels (The Frost Sisters)^, Grace Byers (Reeva Page)\**
\*Regular for season two only. ^Also recurring season one.

**Episodes:**

1.1 eXposed
1.2 rX
1.3 eXodus
1.4 eXit strategy
1.5 boXed in
1.6 got your siX
1.7 eXtreme measures
1.8 threat of eXtinction
1.9 outfoX
1.10 eXploited
1.11 3 X 1
1.12 eXtraction
1.13 X-roads
2.1 eMergence
2.2 unMoored
2.3 coMplications
2.4 outMatched
2.5 afterMath
2.6 iMprint
2.7 no Mercy
2.8 the dreaM
2.9 gaMe changer
2.10 eneMy of My enemy
2.11 meMento
2.12 hoMe
2.13 teMpted
2.14 calaMity
2.15 Monsters
2.16 oMens

**Notable Guest Cast:** *Hayley Lovitt (Sage), Garret Dillahunt (Roderick Campbell),* Joe Nemmers (Ed Weeks), *Jermaine Rivers (Shatter), Elena Satine (Sonya Simonson/Dreamer), Michael Luwoye (Erg),* Anjelica Bette Fellini (Twist), Frances Turner (Paula Turner), Jeff Daniels Phillips (Fade), *Erinn Ruth (Evangeline)*

**Prod:** Neal Ahern; **Music:** John Ottman, David Buckley; **Exec.Prod:** Stan Lee, Alan Fine, Karim Zreik, Joe Quesada, Jim Chory, Jeph Loeb, Len Wiseman, Simon Kinberg, Bryan Singer, Lauren Shuler Donner, Matt Nix; **Created by** Matt Nix; **DOP:** Newton Thomas Sigel; **Prod.Des.**: Derek R Hill; **Costumes:** Louise Mingenbach [1.1], Cameron Dale [1.2-1.13], Karyn Wagner [2.1-2.16]; **Flying Glass of Milk Productions/Donners Company/Bad Hat Harry/Kinberg Genre/Marvel Television/FX Productions**; 45

**Stan Spotting:** Surprisingly, yes, he does pop up in the very first episode to leave a bar as Eclipse enters.

**Notes:** The second of the two television series devised with Marvel for Fox. This series has stronger ties to the X-Men, though it is stated that neither the X-Men nor the Brotherhood may even exist any longer, and that the Sentinel Division of the government is now in operation. Blink's real name has changed from **Days of Future Past** (or, at the very least, she goes by a different surname in this series). The first episode was written by Nix, and directed by

Singer. In April, 2019 – and again probably because of the Disney buyout of Fox – it was confirmed that the program was cancelled.

**Review:** There's a ton of good ideas in *The Gifted*, but their execution is sadly lacking, the performances tend to be quite bland, and many of the good characters are sidelined. Polaris and The Frost Sisters are comfortably the best characters in the show, while mutant hunter Jace Turner is woeful. The Strucker family are annoying, and Blink is cool, but utterly useless. Warpath and Eclipse are also frustrating. The thing is, it's easy to see how the series could have been made engaging, but instead we are given characters that make bizarre decisions, and then rail at the consequences. The cancellation was something of a blessing.

Also Available:

# IT'S ALL CONNECTED

**The Unofficial and Unauthorised Guide to the Marvel
Cinematic Universe**
**The Infinity Saga**
*by Ryan Alcock*

From 2008 until 2019, Marvel Studios produced twenty three films
starring a veritable cavalcade of stars who returned time and again
to build a universe that moviegoers fell in love with. Alongside that,
Marvel Television produced twelve television series that ran on
different networks, exploring more Marvel characters living in the
same universe. It was a monumental media achievement.
But not everyone has the time to watch everything. Not everyone
can pick up on the little details that tie the universe together. Some
people actually have lives. Fortunately, not everyone.
This book is a guide to the Marvel Cinematic Universe in its first
three phases - The Infinity Saga - unofficial and unauthorised,
looking at the details, both nerdy and geeky, that you may have
missed when watching, and supplying the information for the
things you didn't watch.
For both newcomers and established fans, this guide is the perfect
reference work for the MCU.

Also Available:

# IT'S ALL CONNECTED

**The Unofficial and Unauthorised Guide to the Marvel Cinematic Universe**

## Volume Three: 2020 - 2021

### *by Ryan Alcock*

Phase Four of the Marvel Cinematic Universe saw Marvel take a slightly difference approach to what had gone before. With the television series now unde the control of the movie studios, the decision was taken to tie them much closer to the movies. Suddenly the connections were much more important than they had been before. And that was even before they decided to connect to the movies that weren't even part of the MCU. There's just so much to watch...

And not everyone has the time to watch everything. Not everyone can pick up on the little details that tie the universe together. Some people actually have lives. Fortunately, not everyone.

This book is a guide to the Marvel Cinematic Universe in its fourth phase - unofficial and unauthorised, looking at the details, both nerdy and geeky, that you may have missed when watching, and supplying the information for the things you didn't watch.

For both newcomers and established fans, this guide is the perfect reference work for the MCU.